Frontiers of Fractal Analysis

Recent Advances and Challenges

Editors

Santo Banerjee
Dipartimento di Scienze Matematiche
Politecnico di Torino
Corso Duca degli Abruzzi, Torino, Italy

A. Gowrisankar
Department of Mathematics
School of Advanced Sciences
Vellore Institute of Technology, Vellore, India

CRC Press
Taylor & Francis Group
Boca Raton London New York

CRC Press is an imprint of the
Taylor & Francis Group, an **informa** business
A SCIENCE PUBLISHERS BOOK

First edition published 2022
by CRC Press
6000 Broken Sound Parkway NW, Suite 300, Boca Raton, FL 33487-2742

and by CRC Press
2 Park Square, Milton Park, Abingdon, Oxon, OX14 4RN

CRC Press is an imprint of Taylor & Francis Group, LLC

Library of Congress Cataloging-in-Publication Data (applied for)

ISBN: 978-1-032-13867-1 (hbk)
ISBN: 978-1-032-13873-2 (pbk)
ISBN: 978-1-003-23120-2 (ebk)

DOI: 10.1201/9781003231202

Typeset in Times New Roman
by Radiant Productions

Preface

One of the best achievements in human research is the better understanding of natural phenomena that can be represented by mathematical models. The history of describing natural objects using geometry is as old as the advent of science itself. Traditionally lines, squares, rectangles, circles, spheres, etc., have been the basis of our intuitive understanding of the geometry. However, nature is not restricted to such Euclidean objects which are only characterized typically by integer dimensions. Hence, the conventional geometric approach cannot meet the requirements of solving or analysing nonlinear problems which are related to natural phenomena, therefore, fractal theory has been revolutionized by Benoit B Mandelbrot, who aimed to understand complexity and provide an innovative way to recognize irregularity and complex systems. Although the concepts of fractal geometry have found wide applications in many of the forefront areas of science, engineering and societal issues they also have interesting implications of a more practical nature for the older classical areas of science. The idea of fractals, in fact, enables us to see a certain symmetry and order even in an otherwise seemingly disordered and complex system. The importance of the discovery of fractals can hardly be exaggerated. Since its discovery, there has been a surge of research activities in using this powerful concept in almost every branch of scientific disciplines to gain deep insights into many unresolved problems. A large number of applications dealing with the fractal geometry of things as diverse as the price changes and salary distributions, turbulence, statistics of error in telephone messages, word frequencies in written texts, in aggregation and fragmentation processes are just to name a few. The book on Frontiers of Fractal Analysis: Recent Advances and Challenges aims to bring together cutting-edge research works proposing the application of fractals features in both traditional scientific disciplines and in applied fields. This book contains eight chapters that are concisely described as follows.

Approximation theory encompasses a vast area of mathematics. The current context is primarily concerned with the concept of dimension preserving approximation for real-valued multi-variate continuous functions defined on a domain I^q. The first chapter establishes quite a few results similar to the well-known results of multi-variate constrained approximation in terms of dimension preserving approximants. In particular, indication for construction of multi-variate dimension preserving approximants using the concept of fractal interpolation functions is given. In the last part, some multi-valued fractal operators associated with multivariate α-fractal functions are defined and studied.

The second chapter presents an introduction to the fractal interpolation beginning with a global set-up and then extending to a local, a non-stationary, and finally the novel quaternionic setting. Emphasis is placed on the overall perspective with references given to the more specific questions.

The third chapter aims to establish the notion of non-stationary α-fractal operator and establish some approximations and convergence properties. More specifically, the approximations properties of the non-stationary α-fractal polynomials towards a continuous function is discussed. Here a sequence of maps for the non-stationary iterated function systems (IFS) is used. Further, this chapter shows that the proposed method generalizes the existing stationary interpolant in the sense of IFS. The basic properties of this new notion of interpolant are explored, and its box and Hausdorff dimensions are obtained by comparing it to other well-known results. Additionally, using the method of fractal perturbation of a given function, the associated non-stationary fractal operator is constructed and few of its approximation properties are investigated.

In the fourth chapter, fractal calculus in both its local and non-local form is reviewed. The fractal calculus which is called F^{α}-calculus, is a generalization of ordinary calculus that adapted to the fractal sets, curves, and spaces. The non-local fractal derivatives are given the analogue of the Riemann–Liouville and Caputo fractional derivatives. The analogue of the Laplace transform and the Fourier transform in the fractal calculus are suggested. Discrete scale invariance is studied by using the scale transform. Finally, the applications of fractal calculus in classical mechanics, quantum mechanics, and optics are presented.

Fractal calculus is the calculus involving F^{α}-integral and F^{α}-derivative, where $\alpha \in (0, 1]$ is the dimension of the fractal curve. In defining F^{α}-integral and F^{α}-derivative, the mass function and staircase function plays an important role. The fifth chapter discusses the fractal calculus of non-affine fractal interpolation functions namely hidden variable fractal interpolation function and α-fractal function. The fractal integral of the hidden variable fractal interpolation function is examined by predefining the initial conditions. Similarly, by predefining the initial conditions and imposing some necessary conditions on the α-fractal function, its fractal integral is explored.

The sixth chapter investigates the Borel regularity of the relative centered Hausdorff and packing measures. It is shown that the relative packing measure $\mathscr{P}^g$ is Borel regular. The usual spherical measure $\mathscr{U}^g$ is equivalent to the relative Hausdorff measure $\mathscr{H}^g$ in the case where the measures μ and ν satisfy the doubling condition. Moreover, it is proved that this statement is true even without the doubling condition if $q \geq 1$ and $t > 0$ or if $q \leq 0$ and $t \leq 0$ which implies that $\mathscr{H}^g$ is a Borel regular measure.

The multifractal analysis is concerned with the description of irregular measures and functions when the classical analysis does not work. The main goal is the establishment of a multifractal formalism permitting to describe the distribution of the singularities along the support. In its original formulation, the multifractal formalism has been proved to hold for many cases, such as, doubling measures, self-similar functions and measures, and also Gibbs measures. In the seventh chapter, some non-necessary Gibbs vector-valued measures in the framework of the mixed multifractal analysis are concerned. By introducing a gage function in the multifractal formalism, and considering suitable multifractal densities in the framework of relative mixed multifractal analysis, a possible multifractal formalism is proved to hold. Besides, multifractal regularities of the associated multifractal generalizations of the Hausdorff and packing measures are investigated in the new framework.

In the eighth chapter, a novel method is introduced to optimize the selection of order of fractional derivative for detecting edges in images. Before commencing the process, images are enhanced using fuzzy enhancement membership function with variable α in intuitionistic fuzzy domain. The value of α is optimized by comparing visibility of the images by changing α. Then the enhanced image is involved in the edge detection process using fractional derivative of variable order q. By changing q, the edge images are detected and are compared using fuzzy divergence operator. The edge image with minimum fuzzy divergence is the final edge image. The results are compared with existing state of the art techniques. Finally the proposed method is compared graphically with the existing methods with respect to the multifractal dimensional measures. Qualitatively, the proposed technique seems to detect edge in a better way than the compared methods.

Contents

List of Contributors

Agathiyan A

Department of Mathematics, School of Advanced Sciences, Vellore Institute of Technology, Vellore 632 014, Tamil Nadu, India. Email: agathiyanbhc@gmail.com

Agrawal Vishal

Department of Mathematical Sciences, Indian Institute of Technology (BHU), Varanasi, India 221005. Email: vishal.agrawal1992@gmail.com

Ananthi VP

Department of Mathematics, Gobi Arts & Science College, Gobichettipalayam - 638 453, Erode, Tamil Nadu, India. Email: ananthi@gascgobi.ac.in

Ben Mabrouk Anouar

Algebra, Number Theory and Nonlinear Analysis Laboratory LR18ES15, Department of Mathematics, Faculty of Sciences,University of Monastir, 5000 Monastir, Tunisia. Email: anouar.benmabrouk@fsm.rnu.tn

Douzi Zied

Analysis, Probability & Fractals Laboratory LR18ES17, Department of Mathematics, Faculty of Sciences of Monastir, University of Monastir, 5000-Monastir, Tunisia. Email: zied.douzi@fsm.rnu.tn

Easwaramoorthy D

Department of Mathematics, School of Advanced Sciences, Vellore Institute of Technology, Vellore 632 014, Tamil Nadu, India. Email: easandk@gmail.com

Golmankhaneh Alireza Khalili

Department of Physics, Urmia Branch, Islamic Azad University, Urmia, Iran. Email: alirezakhalili2002@yahoo.co.in

Gowrisankar A

Department of Mathematics, School of Advanced Sciences, Vellore Institute of Technology, Vellore 632 014, Tamil Nadu, India. Email: gowrisankargri@gmail.com

Jha Sangita

Department of Mathematics, NIT Rourkela, Rourkela, India 769008. Email: sangitajha285@gmail.com

Massopust Peter R

Centre of Mathematics, Technical University of Munich, Boltzmannstr. 3, 85748 Garching b. Munich, Germany. Email: massopust@ma.tum.de

Pandey Megha

Department of Mathematical Sciences, Indian Institute of Technology (BHU), Varanasi, India 221005. Email: meghapandey1071996@gmail.com

Selmi Bilel

Analysis, Probability & Fractals Laboratory LR18ES17, Department of Mathematics, Faculty of Sciences of Monastir, University of Monastir, 5000-Monastir, Tunisia.
Email: bilel.selmi@fsm.rnu.tn, bilel.selmi@isetgb.rnu.tn

Som Tanmoy

Department of Mathematical Sciences, Indian Institute of Technology (BHU), Varanasi, India 221005.
Email: tsom.apm@iitbhu.ac.in

Thangaraj C

Department of Mathematics, School of Advanced Sciences, Vellore Institute of Technology, Vellore 632 014, Tamil Nadu, India. Email: thangaraj.c2019@vitstudent.ac.in

TMC Priyanka

Department of Mathematics, School of Advanced Sciences, Vellore Institute of Technology, Vellore 632 014, Tamil Nadu, India. Email: priyankamohan195@gmail.com

Verma Saurabh

Department of Applied Sciences, IIIT Allahabad, Prayagraj, India 211015. Email: saurabhverma@iiita.ac.in

Welch Kerri

Faculty at California Institute of Integral Studies, San Francisco, CA, USA. Email: kwelch@ciis.edu

Chapter 1

Some Remarks on Multivariate Fractal Approximation

Megha Pandey*, Vishal Agrawal and Tanmoy Som

1.1 Introduction

The number π is well known in science, it is listed as ratio of a circle's circumference to its diameter. But one might wonder what is the exact value of π. While solving a numerical problem, we use 3.14 or 3.141, etc., in place of π. But all these values are the nearest possible value of its exact value, these values are known as approximate values. In fact there exist a lot of mathematical numbers whose exact value is a hard nut to crack so we use an approximate value instead of exact value. Approximation is used not only for numbers but we can use it for functions also, i.e., we can even approximate a complex function with nearly simpler functions, that is where a vast area in mathematics comes into the picture, broadly known as *Approximation theory*.

Approximation theory, is an emerging subject of mathematics, and a fairly young subject of mathematics since it requires a precise concept of function, which was only developed in the late 18th century. The development of approximation theory, beginning with Leonhard Euler's cartography work in the latter part of the 18th century and continuing with Sergei Bernstein's contributions in the early 20th century towards the establishment of a new area of function theory.

This theory is mainly concerned with finding better approximations for complex functions, using simpler functions. A certain family of functions, such as polynomials or trigonometric polynomials, are used to approximate a given function in approximation theory where the main objective is being able to keep the approximation error under control. Instead, a global, uniform approximation is emphasized, whereas a local, point wise approximation is de-emphasized.

Each human body has an extensive number of complicated structures. For instance, the respiratory tract has a vast network of bronchial branches, the heart has a vast network of arteries, the renal system has a vast network of tubules, and so on. Some physical systems, like buildings, have an extraordinary geometric and functional complexity. Precisely dealing with these occurrences necessitates a mathematical modeling stage. Due to practical limitations, one can not use Euclidean geometry to fix such issues. Only for smooth and regular shapes does this rule apply. Therefore, a point has zero dimensions, a line has one dimension, a plane has two dimensions, and a volume has three dimensions. On the other hand, fractal geometry concentrates on dimensions which are varying between integers. Fractal dimensions, also known as fractal complexity, is the size of the irregular curves. And this level of specificity has significant benefits in the medical profession. While the human body is composed of fractal structures, fractal analysis permits the measurement and simulation of these events. Fractal objects, as a part of applied mathematics, utilizes special non-Euclidean mathematical objects. "Fractal" is commonly used to describe structures that have self-similar features at varying scales. A fractal dimension is a number that represents the amount of information the fractal holds in the space it occupies. Fractals are found everywhere around us, both inside and outside. They are objects

---

Department of Mathematical Sciences, Indian Institute of Technology (BHU), Varanasi, India 221005.
Emails: vishal.agrawal1992@gmail.com; tsom.apm@iitbhu.ac.in
* Corresponding author: meghapandey1071996@gmail.com

in space that may be divided into a never-ending number of similar or identical pieces smaller and smaller. Since it is self-similar, it may be broken down into pieces that are the same as the main part. Some fractal features having smaller structures are perfect duplicates of larger structures.

The fractal dimension, the primary tool for fractal analysis, is used to explain the relevance of fractals to diabetic retinopathy. The pathology of diabetic retinopathy is examined, as is a comparison between the conventional method of disease diagnosis and a new non-invasive method based on fractal geometry. If we take two samples of twenty retinal images, ten normal and ten abnormal, and compare them. This calculates the box counting dimension as a fractal dimension for each sample. This is accomplished with the help of the fractal analysis software Image J.

In numerous areas of Mathematics and elsewhere, Bernstein polynomials have been applied for approximation of functions, including statistical smoothing and the construction and generalization of Bézier curves. Bernstein Polynomials are used to solve the nonlinear Fredholm Integro-Differential Equations and are also utilized in computer graphics, statistics, numerical analysis, p-adic analysis, and solving differential equations.

From the design of new fonts to the construction of mechanical components and assemblies for large-scale industrial design and manufacture, the curves created by applying Bernstein polynomials are used in a variety of applications. By employing the Bernstein polynomials, it is simple to obtain an explicit polynomial representation of Bézier curves.

Following the seminal work of Barnsley [3], Navascués [22, 23] studied the approximation of functions using their fractal counterparts termed as α-fractal functions. In the same vein, Verma and Masspoust [30] recently introduced the notion of dimension preserving approximation. We use dim and $Gr(f)$ respectively to represent fractal dimension and graph of a function of f.

Various concepts of fractal dimensions are available but we cover only those fractal dimensions that are suitable for this section. We only need to mention the Hausdorff dimension, the box dimension, and the packing dimension defined for nonempty subsets of $\mathbb{R}^n$, $n \in \mathbb{N}$, which are denoted by $\dim_H$, $\dim_B$ and $\dim_P$ respectively. To know these fractal dimensions readers are suggested to go through, for instance, [17, 20]. Fractal Functions, Dimensions and their application in the various fields have been broadly discussed in the books and articles published by Banerjee, A Gowrisankar et al., see for instance [7]. Partial integrals and partial derivatives of bivariate fractal interpolation functions (FIFs) have been studied by S. Chandra, S. Abbas proved some results related to mixed Riemann–Liouville fractional integral and derivative related to some iterated function system (IFS), see for instance [9]. Furthermore they discussed the analysis of mixed Weyl-Marchaud fractional derivative and box dimensions in [9]. Approximation by Bernstein polynomials has been established by V. Totik. Interested readers can visit [10] for more details. Agrawal et al. investigated the concept of dimension-preserving approximation for real-valued bivariate continuous functions, defined on a rectangular domain and numerous results are presented. Our work is motivated by [31].

1.1.1 Preliminaries and Notations

For a non-void subset U of $\mathbb{R}^q$, the diameter is defined as

$$|U| = \sup\left\{\|(x_1, \dots, x_q) - (y_1, \dots, y_q)\| : (x_1, \dots, x_q), (y_1, \dots, y_q) \in U\right\}.$$

A countable (or finite) collections of sets, say $\{U_i\}$, will be pronounced as δ-cover of subset $F \subseteq \mathbb{R}^q$, if all the sets of collection has diameter at most δ. Let s be a non-negative real number and $\delta > 0$. Define

$$H_\delta^s(F) = \inf\left\{\sum_{i=1}^{\infty} |U_i|^s : \{U_i\} \text{ is a } \delta-\text{cover of } F\right\}.$$

Definition 1.1 The s-dimensional Hausdorff measure of F is defined by

$$H^s(F) = \lim_{\delta \to 0} H_\delta^s(F).$$

Definition 1.2 Let $F \subseteq \mathbb{R}^q$ and $s \geq 0$. The Hausdorff dimension of F is

$$\dim_H(F) = \inf\left\{s : H^s(F) = 0\right\} = \sup\{s : H^s(F) = \infty\}.$$

Remark 1.1 If $s = \dim_H(F)$, then $H^s(F)$ may be zero or infinite, or may satisfy $0 < H^s(F) < \infty$. A Borel set satisfying this last condition is called an s-set.

Definition 1.3 Let $F \neq \emptyset$ be a bounded subset of $\mathbb{R}^q$ and let $N_\delta(F)$ be the smallest number of sets of diameter at most δ which can cover F. The lower box dimension and upper box dimension of F, respectively, are defined as follows:

$$\underline{\dim}_B(F) = \underline{\lim}_{\delta \to 0} \frac{\log N_\delta(F)}{-\log \delta},$$

and

$$\overline{\dim}_B(F) = \overline{\lim}_{\delta \to 0} \frac{\log N_\delta(F)}{-\log \delta}.$$

If the above two are equal, we call the common value as the box dimension of F. That is,

$$\dim_B(F) = \lim_{\delta \to 0} \frac{\log N_\delta(F)}{-\log \delta}.$$

The following relations are established between these fractal dimensions (see [17]):

$$\dim_H F \leq \underline{\dim}_B F \leq \overline{\dim}_B F,$$

and

$$\dim_H F \leq \dim_P F \leq \overline{\dim}_B F.$$

Lemma 1.1 ([30], Lemma 3.1)
Let $A \subset \mathbb{R}^m$ and $f, g : A \to \mathbb{R}^n$ be continuous functions. Then,

$$\dim_H(Gr(f+g)) = \dim_H(Gr(g)) \quad \text{and} \quad \dim_P(Gr(f+g)) = \dim_P(Gr(g))$$

provided that f is a Lipschitz function.

Remark 1.2 Note that the above lemma is also true for box dimensions.

Theorem 1.1 ([13], Theorem 9.40)
Suppose f is defined in an open set $E \subset \mathbb{R}^2$, and $D_1 f$ and $D_{21} f$ exists at every point of E. Suppose $Q \subset E$ is a closed rectangle with sides parallel to the coordinate axes, having (a, b) and $(a+h, b+k)$ as opposite vertices $(h \neq 0, k \neq 0)$. Put

$$\Delta(f, Q) = f(a+h, b+k) - f(a+h, b) - f(a, b+k) + f(a, b).$$

Then, there is a point (x, y) in the interior of Q such that

$$\Delta(f, Q) = hk(D_{21}f)(x, y).$$

Where $D_1 f = \frac{\partial f}{\partial x}$, $D_2 f = \frac{\partial f}{\partial y}$, and $D_{21} f = D_2(D_1 f)$.

Note 1.1 The above theorem is known as Mean value theorem. For simplicity (and without loss of generality) the above theorem has been stated for the function of two variables.

1.1.2 Motivation

The class of all real-valued continuous functions on $I^q := \underbrace{I \times \cdots \times I}_{\text{q-times}}$ is defined by $C(I^q)$ where $I = [0, 1]$.

$D^{(k_1,\ldots,k_q)} f$ denotes the derivative of the $(k_1, \ldots, k_q)$-th order of a multi-variate function f, that is, $D^{k_1,\ldots,k_q} f := \dfrac{\partial^{k_1+\cdots+k_q} f}{\partial x_1^{k_1} \ldots \partial x_q^{k_q}}$. Let

$$C^{n_1,\ldots,n_q}(I^q) = \{f : I^q \to \mathbb{R};\ D^{k_1,\ldots,k_q} f \in C(I^q),\ \text{ for all } 0 \le k_i \le n_i\}.$$

If $D^{k_1,\ldots,k_q} f(x_1, \ldots, x_q) \ge 0$, for all $x_1, \ldots, x_q \in I^q$, then we say the function f is $(n_1, \ldots, n_q)$-convex. Let $g \in C(I^q)$ such that $\dim(Gr(g)) > q$. We may refer to [11] for the existence of such functions.

The function $f : I^q \to \mathbb{R}$ defined by $f(x_1, \ldots, x_q) := \int_0^{x_1} \cdots \int_0^{x_q} g(t_1, \ldots, t_q) dt_1\, dt_2 \ldots dt_q$ satisfies the following:

$$\dim(Gr(f)) = q \quad \text{and} \quad \dim Gr(D^{(1,\ldots,1)} f) = \dim(Gr(g)) > q,$$

where dim denotes a fractal dimension.

Recall that the tensor product Bernstein polynomial on I^q is defined as:

$$B_{m_1,\ldots,m_q}(f)(x_1, \ldots, x_q) = \sum_{i_1=0}^{m_1} \cdots \sum_{i_q=0}^{m_q} f\left(\frac{i_1}{m_1}, \cdots, \frac{i_q}{m_q}\right) \binom{m_1}{i_1} \cdots \binom{m_q}{i_q} x_1^{i_1}(1 - x_1)^{m_1 - i_1} \ldots x_q^{i_q}(1 - x_q)^{m_q - i_q}.$$

If we approximate a function $f \in C^{k_1,\ldots,k_q}(I^q)$ with $B_{m_1,\ldots,m_q}(f)$, we obtain the following (see [12] for a discussion of numerous Bernstein polynomial properties).

- $B_{m_1,\ldots,m_q}(f) \to f$ uniformly on I^q.
- $\left(D^{k_1,\ldots,k_q}(B_{m_1,\ldots,m_q}(f))\right) \to D^{k_1,\ldots,k_q} f$ uniformly on I^q.
- Since $B_{m_1,\ldots,m_q}(f)$ and $D^{k_1,\ldots,k_q}(B_{m_1,\ldots,m_q}(f))$ are polynomials, then

$$\dim\left(Gr(D^{k_1\ldots,k_q}(B_{m_1,\ldots,m_q}(f)))\right) = \dim(Gr(B_{m_1,\ldots,m_q}(f))) = \dim(Gr(f)) = q.$$

The preceding statements may lead to the conclusion that the Bernstein polynomial approximation preserves the smoothness of a function, however these don't guarantee for the dimensions of its partial derivatives.

Hence the motivation of the current study is to examine the viewpoint of approximation concerning a function's fractal dimension and its partial derivatives.

The next Section 1.2 is devoted to building the basic theories which will be required to develop dimension preserving approximation of multi-variate functions which has been introduced in Subsection 1.2.1. The chapter has been concluded in Section 1.3 followed by the references. In this last section we have constructed and established various properties for multi-valued mappings defined using multi-variate α-fractal functions.

1.2 Dimension Preserving Approximation of Multivariate Functions

Let us denote the class of real-valued Lipschitz functions on I^q by $\mathcal{L}ip(I^q)$, note that this space is a dense subset of $C(I^q)$ with respect to the supremum norm.

In view of Lipschitz invariance property of dimension, one may conclude that the upcoming theorem holds for all aforementioned dimensions.

Theorem 1.2 *Let $q \le \beta \le q+1$. Then, the set $\mathcal{S}_\beta := \{f \in C(I^q) : \dim(Gr(f)) = \beta\}$ is dense in $C(I^q)$.*

Proof Let $f \in C(I^q)$ and $\epsilon > 0$. Using the density of $\mathcal{L}ip(I^q)$ in $C(I^q)$, there exists g in $\mathcal{L}ip(I^q)$ such that

$$\|f - g\|_\infty < \frac{\epsilon}{2}.$$

Further, we consider a non-vanishing function $h \in \mathcal{S}_\beta$. Let $h_* = g + \frac{\epsilon}{2\|h\|_\infty} h$, which immediately gives

$$\|g - h_*\|_\infty \le \frac{\epsilon}{2}.$$

This together with Lemma 1.1 it implies that $\dim(Gr(h_*)) = \dim(Gr(h)) = \beta$. Hence, we have $h_* \in \mathcal{S}_\beta$ and

$$\|f - h_*\|_\infty \le \|f - g\|_\infty + \|g - h_*\|_\infty < \epsilon.$$

Thus, the proof of the theorem is complete. □

The next theorem seems to be widely known in the univariate context, however, we were unable to discover the proof in the multi-variate scenario. thus, for completeness, we include the proof of the following theorem.

Theorem 1.3 *Let (f_k) be a sequence of differentiable functions on I^q. Assume that for some $(w_1, \dots, w_q) \in I^q$, the sequences $(f_k(w_1, \dots))$, $(f_k(\cdot, w_2, \dots), \dots$ and $(f_k(\dots, w_q))$ converge uniformly on I^{q-1}. If $(D^{(1,\dots,1)} f_k)$ converges uniformly on I^q, then (f_k) converges uniformly on I^q to a function f, and*

$$D^{(1,\dots,1)} f(x_1, \dots, x_q) = \lim_{k\to\infty} D^{(1,\dots,1)} f_k(x_1, \dots, x_q),$$

for every $x_1, \dots, x_q \in I^q$.

Proof Let $\epsilon > 0$. Since $(D^{(1,\dots,1)} f_k)$ converges uniformly, there exists $N_1 \in \mathbb{N}$ such that

$$|D^{(1,\dots,1)} f_k(x_1, \dots, x_q) - D^{(1,\dots,1)} f_m(x_1, \dots, x_q)| < \frac{\epsilon}{2^q}, \quad \text{for all } x_1, \dots, x_q \in I^q,\ k, m \ge N_1.$$

Applying Theorem 1.1 on $f_k - f_m$, we have

$$\begin{aligned}
&\Big|\big[f_k(x_1+h_1,\ldots,x_q+h_q) - f_m(x_1+h_1,\ldots,x_q+h_q)\big]\\
&\quad - \sum_{1\le i\le q}\big[f_k(x_1+h_1,\ldots,x_{i-1}+h_{i-1},x_i,x_{i+1}+h_{i+1},\ldots,x_q+h_q)\\
&\qquad\qquad - f_m(x_1+h_1,\ldots,x_{i-1}+h_{i-1},x_i,x_{i+1}+h_{i+1},\ldots,x_q+h_q)\big]\\
&\quad + \sum_{1\le i<j\le q}\big[f_k(x_1+h_1,\ldots,x_{i-1}+h_{i-1},x_i,x_{i+1}+h_{i+1},\ldots,x_{j-1}+h_{j-1},x_j,x_{j+1}+h_{j+1},\\
&\qquad \ldots,x_q+h_q) - f_m(x_1+h_1,\ldots,x_{i-1}+h_{i-1},x_i,x_{i+1}+h_{i+1},\ldots,x_{j-1}+h_{j-1},x_j,\\
&\qquad\qquad x_{j+1}+h_{j+1},\ldots,x_q+h_q)\big] + \cdots + (-1)^q\big[f_k(x_1,\ldots,x_q) - f_m(x_1,\ldots,x_q)\big]\Big|\\
&= h_1\ldots h_q\left|D^{(1,\ldots,1)}(f_k-f_m)(t_1,\ldots,t_q)\right|\\
&\le h_1\ldots h_q \max_{(t_1,t_2,\ldots,t_q)\in I^q}\left|D^{(1,\ldots,1)}f_k(t_1,\ldots,t_q) - D^{(1,\ldots,1)}f_m(t_1,\ldots,t_q)\right|\\
&< \frac{\epsilon}{2^q}h_1\ldots h_q\\
&< \frac{\epsilon}{2^q}.
\end{aligned}\tag{1.1}$$

Using the assumption for $(w_1,\ldots,w_q)\in I^q$, we can choose $N_0\ (>N_1)\ \in\mathbb{N}$ such that

$$\begin{aligned}
&|f_k(w_1,x_2,x_3\ldots,x_q) - f_m(w_1,x_2,x_3\ldots,x_q)| < \frac{\epsilon}{2^q} \quad \text{for all } k,m\ge N_0,\\
&|f_k(x_1,w_2,x_3\ldots,x_q) - f_m(x_1,w_2,x_3\ldots,x_q)| < \frac{\epsilon}{2^q} \quad \text{for all } k,m\ge N_0,\\
&\vdots\\
&|f_k(x_1,\ldots,x_{q-1},w_q) - f_m(x_1,\ldots,x_{q-1},w_q)| < \frac{\epsilon}{2^q} \quad \text{for all } k,m\ge N_0.
\end{aligned}$$

Now, using the above estimates and Equation (1.1) we have

$$\begin{aligned}
|f_k(x_1,x_2,\ldots,x_q) - f_m(x_1,x_2,\ldots,x_q)| &< \frac{\epsilon}{2^q} + |f_k(w_1,x_2,\ldots,x_q) - f_m(w_1,x_2,\ldots,x_q)| +\\
&\quad \cdots + |f_k(x_1,\ldots,x_{q-1},w_q) - f_m(x_1,\ldots,x_{q-1},w_q)|\\
&< \frac{\epsilon}{2^q} + \cdots + \frac{\epsilon}{2^q}\\
&< \epsilon,
\end{aligned}$$

for every $(x_1,\ldots,x_q)\in I^q$ and $k,m\ge N_0$. This immediately confirms the uniform convergence of (f_k). The rest follows by routine calculations, and is hence omitted. □

Lemma 1.2 *Let $f: I^{q-1}\to\mathbb{R}$ be a Lipschitz map with lipschitz constant L and $g: I\to\mathbb{R}$ be a continuous function. A mapping $h: I^q\to\mathbb{R}$ defined by*

$$h(x_1,\ldots,x_q) = f(x_1,\ldots,x_{q-1}) + g(x_q).$$

Then, $\dim_H(Gr(h)) = \dim_H(Gr(g)) + q - 1$.

Proof Define a map $\phi : I^{q-1} \times Gr(g) \to Gr(h)$ such that

$$\phi(x_1, \dots, x_q, g(x_q)) = (x_1, \dots, x_q, h(x_1, \dots, x_q)),$$

where $(x_1, \dots, x_q) = (x_1, \dots, x_q)$.

Claim: ϕ is a bi-Lipschitz mapping. For this, take $M = \max\left\{\sqrt{1 + 2L^2}, \sqrt{2}\right\}$, then for all $(x_1, \dots, x_q)$, $(y_1, \dots, y_q) \in I^q$, we have

$$\begin{aligned}
&\left\|\phi((x_1, \dots, x_q, g(x_q)) - \phi(y_1, \dots, y_q, g(y_q))\right\|_2 \\
&= \left\|(x_1, \dots, x_q, h(x_1, \dots, x_q)) - (y_1, \dots, y_q, h(y_1, \dots, y_q))\right\|_2 \\
&= \sqrt{\|(x_1, \dots, x_q) - (y_1, \dots, y_q)\|_2^2 + (h(x_1, \dots, x_q) - h(y_1, \dots, y_q))^2} \\
&= \sqrt{\|(x_1, \dots, x_q) - (y_1, \dots, y_q)\|_2^2 + (f(x_1, \dots, x_q) + g(x_q) - f(y_1, \dots, y_q) - g(y_q))^2} \\
&\le \sqrt{\|(x_1, \dots, x_q) - (y_1, \dots, y_q)\|_2^2 + 2(f(x_1, \dots, x_q) - f(y_1, \dots, y_q))^2 + 2(g(x_q) - g(y_q))^2} \\
&\le \sqrt{\|(x_1, \dots, x_q) - (y_1, \dots, y_q)\|_2^2 + 2L^2\|(x_1, \dots, x_q) - (y_1, \dots, y_q)\|_2^2 + 2(g(x_q) - g(y_q))^2} \\
&= \sqrt{(1 + 2L^2)\|(x_1, \dots, x_q) - (y_1, \dots, y_q)\|_2^2 + 2(g(x_q) - g(y_q))^2} \\
&= M\sqrt{\|(x_1, \dots, x_q) - (y_1, \dots, y_q)\|_2^2 + 2(g(x_q) - g(y_q))^2} \\
&= M\|(x_1, \dots, x_q, g(x_q)) - (y_1 \dots, y_q, g(y_q))\|_2.
\end{aligned}$$

Hence,

$$\left\|\phi((x_1, \dots, x_q, g(x_q)) - \phi(y_1, \dots, y_q, g(y_q))\right\|_2 \le M\|(x_1, \dots, x_q, g(x_q)) - (y_1 \dots, y_q, g(y_q))\|_2.$$

With small manipulation, we can similarly prove that

$$\left\|\phi((x_1, \dots, x_q, g(x_q)) - \phi(y_1, \dots, y_q, g(y_q))\right\|_2 \ge \frac{1}{M}\|(x_1, \dots, x_q, g(x_q)) - (y_1 \dots, y_q, g(y_q))\|_2.$$

This proves the claim. Therefore, $\dim_H(Gr(h)) = \dim_H(I^{q-1} \times Gr(g))$, but I^{q-1} and $Gr(g)$ both are Borel sets and that is why $\dim_H(I^{q-1} \times Gr(g)) = \dim_H(I^{q-1}) + \dim_H(Gr(g))$
$= \dim_H(Gr(g)) + q - 1$, which proves our lemma.

Here, we shall discuss the results obtained for univariate functions concerning dimensional aspects. One important class of functions considered by Mauldin and Williams [14] is as follows:

$$W_b(x) := \sum_{n=-\infty}^{\infty} b^{-\alpha n}[\phi(b^n x + \theta_n) - \phi(\theta_n)],$$

where θ_n is an arbitrary real number, ϕ is a periodic function with period one and $b > 1$, $0 < \alpha < 1$. They showed that for a large enough b there exists a constant $C > 0$ such that $\dim_H(Gr(W_b)$ is bounded below by $2 - \alpha - (C/\ln b)$. Further, a significant progress in dimension theory of functions is contributed by Shen [11] for the following class of functions:

$$f^{\phi}_{\lambda,b}(x) := \sum_{n=0}^{\infty} \lambda^n \phi(b^n x),$$

where $b \geq 2$ and ϕ is a real-valued, $\mathbb{Z}$-periodic, non-constant, C^2-function defined on $\mathbb{R}$. Shen proved that there exists a constant K_0 depending on ϕ and b such that if $1 < \lambda b < K_0$, then

$$\dim_H(Gr(f^{\phi}_{\lambda,b})) = 2 + \frac{\log \lambda}{\log b}.$$

For $f \in C^{1,\ldots,1}(I^q)$, we get $\dim(Gr(f)) = q$. However, no conclusion can be drawn for dimensions of its partial derivatives. This is evident from the following example: let Weierstrass-type nowhere differentiable continuous function $W : I \to \mathbb{R}$ as in [11] with $1 \leq \dim(Gr(W)) \leq 2$. Now, we define $h : I^q \to \mathbb{R}$ by

$$h(x_1, \ldots, x_q) = W(x_q) + x_1 + \cdots + x_{q-1}.$$

Here, by Lemma 1.2, we obtain $q \leq \dim(Gr(h)) = \dim(Gr(W)) + q - 1 \leq q + 1$. Then, for the function f defined by

$$f(x_1, \ldots, x_q) := \int_0^{x_1} \cdots \int_0^{x_q} h(t_1, \ldots, t_q)\, dt_1 \ldots dt_q,$$

we have $\dim(Gr(f)) = q$ and $q \leq \dim(Gr(D^{(1,\ldots,1)} f)) = \dim(Gr(h)) \leq q + 1$.

Theorem 1.4 *Let $f \in C^{1,\ldots,1}(I^q)$ such that $\dim(Gr(D^{(1,\ldots,1)} f)) = \beta$ for some $q \leq \beta \leq q + 1$. Then, we have a sequence (f_k) in $C^{1,\ldots,1}(I^q)$ such that $\dim(Gr(D^{(1,\ldots,1)} f_k)) = \beta$ and $f_k \to f$ uniformly on I^q.*

Proof In view of Theorem 1.2, there exists a sequence (g_k) in $C(I^q)$ such that $\dim(Gr(g_k)) = \beta$ and $g_k \to D^{(1,\ldots,1)} f$ uniformly on I^q. Further, let us consider a function $f_k : I^q \to \mathbb{R}$ defined by

$$f_k(x_1, \ldots, x_q) := \int_0^{x_1} \cdots \int_0^{x_q} g_k(t_1, \ldots, t_q) dt_1 \ldots dt_q.$$

Then, $D^{(1,\ldots,1)} f_k = g_k$ and $(D^{(1,\ldots,1)} f_k) \to D^{(1,\ldots,1)} f$ uniformly. Next, we have that the sequences of functions $\big(f_k(0, y_2, \ldots, y_q)\big) \to 0, \ldots, \big(f_k(y_1, \ldots, y_{q-1}, 0)\big) \to 0$, uniformly on I^{q-1}. Now, Theorem 1.3 completes the proof. □

Theorem 1.5 *Let $f \in C(I^q)$ with $f(x_1, \ldots, x_q) \geq 0$ for all $(x_1, \ldots, x_q) \in I^q$. Then, for a given $\epsilon > 0$, there exists $g \in \mathcal{S}_\beta$ satisfying the following:*

$$g(x_1, \ldots, x_q) \geq 0 \text{ for all } (x_1, \ldots, x_q) \in I^q \text{ and } \|f - g\|_\infty < \epsilon.$$

Proof Let $\epsilon > 0$. Theorem 1.2 yields an element $h \in \mathcal{S}_\beta$ such that

$$\|f - h\|_\infty < \frac{\epsilon}{2}.$$

We define

$$g(x_1, \ldots, x_q) := h(x_1, \ldots, x_q) + \frac{\epsilon}{2}, \text{ for all } x_1, \ldots, x_q \in I^q.$$

Then, by Lemma 1.1, $g \in \mathcal{S}_\beta$, and by routine calculations, we get

$$\begin{aligned} g(x_1,\dots,x_q) &= h(x_1,\dots,x_q) - f(x_1,\dots,x_q) + f(x_1,\dots,x_q) + \frac{\epsilon}{2} \\ &\geq -\|f-h\|_\infty + f(x_1,\dots,x_q) + \frac{\epsilon}{2} > f(x_1,\dots,x_q) \geq 0. \end{aligned}$$

Furthermore, one has

$$\|f-g\|_\infty \leq \|f-h\|_\infty + \|h-g\|_\infty < \epsilon,$$

hence the proof. □

Theorem 1.6 *Let $f : I^q \to \mathbb{R}$ be a $(n_1,\dots,n_q)$-convex function such that $f(0,x_2,\dots,x_q) = \cdots = f(x_1,\cdots,x_{q-1},0) = 0$, for all $x_1,\dots,x_q \in I$. Then, for $\epsilon > 0$, there exists $(n_1,\dots,n_q)$-convex function g such that $D^{(n_1,\dots,n_q)}g \in \mathcal{S}_\beta$ and $\|f-g\|_\infty < \epsilon$.*

Proof Let $\epsilon > 0$. Using Theorem 1.2, there exists $h \in \mathcal{S}_\beta$ such that $\|D^{(n_1,\dots,n_q)}f - h\| < \epsilon$. By choosing

$$g(x_1,\dots,x_q) := \int_0^{x_1}\cdots\int_0^{x_q}\cdots\int_0^{x_1^{n_1-1}}\cdots\int_0^{x_q^{n_q-1}} h(x_1^{n_1},\dots,x_q^{n_q})dx_1^{n_1}\dots dx_q^{n_q}\dots dx_1^1\dots dx_q^1$$

we have

$$\begin{aligned} &\|f-g\|_\infty \\ &= \sup\left\{\left|f - \int_0^{x_1}\cdots\int_0^{x_q}\cdots\int_0^{x_1^{n_1-1}}\cdots\int_0^{x_q^{n_q-1}} h(x_1^{n_1},\dots,x_q^{n_q})dx_1^{n_1}\dots dx_q^{n_q}\dots dx_1^1\dots dx_q^1\right| \right. \\ &\qquad\qquad\qquad\qquad \left. : (x_1,\dots,x_q) \in I^q\right\} \\ &< \epsilon, \end{aligned}$$

proving the assertion. □

Theorem 1.7 *Let $f \in C(I^q)$. Then, for $\epsilon > 0$ there exists $g \in \mathcal{S}_\beta$ such that*

$$g(x_1,\dots,x_q) \leq f(x_1,\dots,x_q) \text{ for all } x_1,\dots,x_q \in I^q \text{ and } \|f-g\|_\infty < \epsilon.$$

Proof Since $f \in C(I^q)$ and $\epsilon > 0$, Theorem 1.2 generates a member $h \in \mathcal{S}_\beta$ such that

$$\|f-h\|_\infty < \frac{\epsilon}{2}.$$

Choose $g(x_1,\dots,x_q) := h(x_1,\dots,x_q) - \frac{\epsilon}{2}$, for all $x_1,\dots,x_q \in I^q$. Then,

$$\begin{aligned} g(x_1,\dots,x_q) &= h(x_1,\dots,x_q) - f(x_1,\dots,x_q) + f(x_1,\dots,x_q) - \frac{\epsilon}{2} \\ &\leq \|f-h\|_\infty + f(x_1,\dots,x_q) - \frac{\epsilon}{2} < f(x_1,\dots,x_q). \end{aligned}$$

Furthermore,

$$\|f-g\|_\infty \le \|f-h\|_\infty + \|h-g\|_\infty < \epsilon,$$

establishing the proof. □

We now intend to demonstrate the existence of best one-sided approximation. Consider $\beta \in [q, q+1]$, and define

$$\mathcal{C}_\beta(I^q) := \{f \in \mathcal{C}(I^q) : \overline{\dim}_B(Gr(f)) \le \beta\}.$$

In view of [15, Proposition 3.4], recall that $\mathcal{C}_\beta(I^q)$ is a normed linear space. Let $\{g_1, \ldots, g_n\}$ be a linearly independent subset of $\mathcal{C}_\beta(I^q)$. Further, for a bounded below and Lebesgue integrable function $f : I^q \to \mathbb{R}$, we define

$$\mathcal{Y}_n^\beta(f) := \Big\{h \in span\{g_1, \ldots, g_n\} : h(x_1, \ldots, x_q) \le f(x_1, \ldots, x_q) \text{ for all } (x_1, \ldots, x_q) \in I^q\Big\}.$$

Theorem 1.7 ensures that $\mathcal{Y}_n^\beta(f)$ is non-empty. A function $h_f \in \mathcal{Y}_n^\beta(f)$ is said to be a best one-sided approximation from below to f on I^q if

$$\int_{I^q} h_f(x_1, \ldots, x_q)\, dx_1 \ldots dx_q = \sup\left\{\int_{I^q} h(x_1, \ldots, x_q)\, dx_1 \ldots dx_q : h \in \mathcal{Y}_n^\beta(f)\right\},$$

where $\displaystyle\int_{I^q} h_m(x_1, \ldots, x_q)\, dx_1 \ldots dx_q = \underbrace{\int_0^1 \cdots \int_0^1}_{\text{q-times}} h_m(x_1, \ldots, x_q)\, dx_1 \ldots dx_q.$

In a similar way, we define the best one-sided approximations from above. We state the next theorem for one-sided approximation from below. Though a similar result can be proved in terms of one-sided approximation from above, see, for instance, [16, 29].

Theorem 1.8 *For a bounded below and integrable function $f : I^q \to \mathbb{R}$, there exists a member in $\mathcal{Y}_n(f)$ of best one-sided approximant from below to f on I^q.*

Proof Let (h_m) be a sequence in $\mathcal{Y}_n(f)$ such that

$$\int_{I^q} h_m(x_1, \ldots, x_q)\, dx_1 \ldots dx_q \to A \quad \text{as } m \to \infty, \tag{1.2}$$

where

$$\int_{I^q} h_m(x_1, \ldots, x_q)\, dx_1 \ldots dx_q = \int_0^1 \cdots \int_0^1 h_m(x_1, \ldots, x_q)\, dx_1 \ldots dx_q,$$
$$\text{and } A = \sup\left\{\int_{I^q} h(x_1, \ldots, x_q)\, dx_1 \ldots dx_q : h \in \mathcal{Y}_n^\beta(f)\right\}.$$

With an appropriate constant $M_* > 0$, we have

$$\int_{I^q} |h_m(x_1, \ldots, x_q)|\, dx_1 \ldots dx_q \le \int_{I^q} \Big|h_m(x_1, \ldots, x_q) - A\Big|\, dx_1 \ldots dx_q + \int_{I^q} A\, dx_1 \ldots dx_q \le M_*.$$

Since $\mathcal{Y}_n^\beta(f)$ is a subset of finite-dimensional linear space, the closed set of radius M_* in $\mathcal{Y}_n^\beta(f)$ is compact. Therefore, there exist a subsequence (h_{m_k}) and a function h in $\mathcal{Y}_n^\beta(f)$ such that the sequence (h_{m_k}) converges to h in $\mathcal{L}^1(I^q)$. By using the fact that every norm is equivalent on a finite-dimensional linear space. Now, from the finite-dimensionality of $\mathcal{Y}_n^\beta(f)$, it follows that the sequence (h_{m_k}) also converges to h uniformly. Further, since $h_m(x_1,\dots,x_q) \le f(x_1,\dots,x_q)$, for all $(x_1,\dots,x_q) \in I^q$, and $h_{m_k} \to h$ uniformly, we get $h(x_1,\dots,x_q) \le f(x_1,\dots,x_q)$, for all $(x_1,\dots,x_q) \in I^q$. Thus, $h \in \mathcal{Y}_n^\beta(f)$. Now, by (1.2), we have

$$\int_{I^q} h(x_1,\dots,x_q)\, dx_1 \dots dx_q = \lim_{k\to\infty} \int_{I^q} h_{m_k}(x_1,\dots,x_q)\, dx_1 \dots dx_q = A,$$

proving the assertion. □

1.2.1 Construction of Dimension Preserving Approximants

First, Hutchinson [13] hinted at the generation of parameterized fractal curves. In [3], Barnsley introduced Fractal Interpolation Functions (FIFs) via Iterated Function System (IFSs). It is important to choose the IFS appropriately so that it is fitted as an attractor for a graph of a continuous function called FIF. We refer to the reader [3] for more study regarding the construction of FIFs.

Computation of dimensions of fractal functions has been an integral part of fractal geometry. In [3], Barnsley has proved estimates for the Hausdorff dimension of an affine FIF. Falconer also established a similar result in [19]. Barnsley and his collaborators [20–22] have computed the box dimension of classes of affine FIFs. In [20], FIFs generated by bilinear maps have been studied. In [23], a formula for the box dimension of FIFs $\mathbb{R}^n \to \mathbb{R}^m$ has been proved. A particular case of FIFs given by Navascués [22], namely, (univariate) *α-fractal function* has been proven very useful in approximation theory and operator theory. Using series expansion, the box dimension of (univariate) α-fractal function is estimated in [24]. Following the work of Ruan and Xu [25], Verma and Viswanathan [28] introduced bivariate version of α-fractal function and studied their applications in approximation theory.

Here we plan to provide a construction of multi-variate Bernstein α-fractal function, which is motivated by Verma and Viswanathan [28] on rectangular grids.

During the preparation of this manuscript, we have noticed a similar construction of multivariate α-fractal function in [27]. However, it is worth noting that our aim is different from that of [27]. In particular, we focus on fractal dimensions of multivariate α-fractal functions.

Let $f \in C(I^q)$. Let us denote $\Sigma_k = \{1,2,\dots,k\}$, $\Sigma_{k,0} = \{0,1,\dots k\}$, $\partial\Sigma_{k,0} = \{0,k\}$ and $\text{int}\Sigma_{k,0} = \{1,2,\dots,k-1\}$. Further, a net Δ on I^q is defined as follows:

$$\Delta := \left\{(x_{1,i_1},\dots,x_{q,i_q}) \in I^q : i_k \in \Sigma_{N_k,0},\ 0 = x_{k,0} < \dots < x_{k,N_k} = 1,\ k \in \Sigma_q\right\}.$$

Note that $\{x_{0,k},\dots,x_{k,N_k}\}$ forms a partition of the interval I_k, where I_k denotes k-th unit interval in the product I^q, with the aid of which Δ forms a partition of I^q. For each $i_k \in \Sigma_{N_k}$, let us define $I_{k,i_k} = [x_{k,i_k-1}, x_{k,i_k}]$ and linear contraction mappings $u_{k,i_k} : I_k \to I_{k,i_k}$ such that

$$u_{k,i_k}(x_{k,0}) = x_{k,i_k-1},\quad u_{k,i_k}(x_{k,N_k}) = x_{k,i_k},\quad \text{if } i_k \text{ is odd, and}$$

$$u_{k,i_k}(x_{k,0}) = x_{k,i_k},\quad u_{k,i_k}(x_{k,N_k}) = x_{k,i_k-1},\quad \text{if } i_k \text{ is even.}$$

For each, let us denote $Q_{i_1\dots i_q}(x_1,\dots,x_q) := \left(u_{1,i_1}^{-1}(x_1),\dots,u_{q,i_q}^{-1}(x_q)\right)$, where $(x_1,\dots,x_q) \in \prod_{k=1}^q I_{k,i_k}$.

Let $\alpha \in C(I^q)$ be such that $\|\alpha\|_\infty < 1$. Assume further that $s \in C(I^q)$ satisfies $s(x_{1,j_1},\dots,x_{q,j_q}) = f(x_{1,j_1},\dots,x_{q,j_q})$, for all $(j_1,\dots,j_q) \in \prod_{k=1}^{q} \partial\Sigma_{N_k,0}$. By [27, Theorem 2.1], we have a unique function $f^\alpha_{\Delta,s} \in C(I^q)$ termed as α-fractal function, such that

$$f^\alpha_{\Delta,s}(x_1,\dots,x_q) = f(x_1,\dots,x_q) + \alpha(x_1,\dots,x_q)\, f^\alpha_{\Delta,s}\big(Q_{i_1\dots i_q}(x_1,\dots,x_q)\big)$$
$$-\alpha(x_1,\dots,x_q)\, s\big(Q_{i_1\dots i_q}(x_1,\dots,x_q)\big),$$

for $(x_1,\dots,x_q) \in \prod_{k=1}^{q} I_{k,i_k},\ i_k \in \Sigma_{N_k},\ k \in \Sigma_q$. For our convenience, let us write the metric

$$d_{I^q}((x_1,\dots,x_q),(y_1,\dots,y_q)) := \sqrt{(x_1-y_1)^2+\dots+(x_q-y_q)^2},$$

where $(x_1,\dots,x_q),\ (y_1,\dots,y_q) \in I^q$.

Definition 1.4 A function $f : I^q \to \mathbb{R}$ is said to be Hölder continuous with exponent σ and Hölder constant K_f if

$$|f(x_1,\dots,x_q) - f(y_1,\dots,y_q)| \le K_f d_{I^q}((x_1,\dots,x_q),(y_1,\dots,y_q))^\sigma,$$

for all $(x_1,\dots,x_q),(y_1,\dots,y_q) \in I^q$, and for some $K_f > 0$.
We define the Hölder space

$$\mathcal{H}^\sigma(I^q) := \{g : I^q \to \mathbb{R} :\ \text{g is Hölder continuous with exponent } \sigma\}.$$

If we equip the space $\mathcal{H}^\sigma(I^q)$ with norm $\|g\| := \|g\|_\infty + [g]_\sigma$ where

$$[g]_\sigma = \sup_{(x_1,\dots,x_q)\neq(y_1,\dots,y_q)} \frac{|g(x_1,\dots,x_q) - g(y_1,\dots,y_q)|}{d_{I^q}((x_1,\dots,x_q),(y_1,\dots,y_q))^\sigma},$$

then it forms a Banach space.

Theorem 1.9 *Let f and s be Hölder continuous with exponent σ such that $s(x_{1,j_1},\dots,x_{q,j_q}) = f(x_{1,j_1},\dots,x_{q,j_q})$, for all $(j_1,\dots,j_q) \in \prod_{k=1}^{q} \partial\Sigma_{N_k,0}$. Further, we assume that $N_1 = \dots = N_q = N$, $x_{k,i_k} - x_{k,i_k-1} = \frac{1}{N}$, for all $i_k \in \Sigma_{N_k},\ k \in \Sigma_q$ and constant scaling function α. If $N^\sigma|\alpha| < 1$, then $f^\alpha_{\Delta,s}$ is Hölder continuous with exponent σ.*

Proof Let $\mathcal{H}^\sigma_f(I^q) = \{g \in \mathcal{H}^\sigma(I^q) : g(x_{1,j_1},\dots,x_{q,j_q}) = f(x_{1,j_1},\dots,x_{q,j_q}),$ for all $(j_1,\dots,j_q) \in \prod_{k=1}^{q} \partial\Sigma_{N_k,0}\}$. We observe that the space $\mathcal{H}^\sigma_f(I^q)$ is a closed subset of $\mathcal{H}^\sigma(I^q)$. It follows that $\mathcal{H}^\sigma_f(I^q)$ is a complete metric space with respect to the metric induced by norm $\|.\|$. We define a map $T : \mathcal{H}^\sigma_f(I^q) \to \mathcal{H}^\sigma_f(I^q)$ by

$$(T_f g)(x_1,\dots,x_q) = f(x_1,\dots,x_q) + \alpha\,(g-s)(Q_{i_1\dots i_q}(x_1,\dots,x_q))$$

for all $(x_1,\dots,x_q) \in \prod_{k=1}^{q} I_{k,i_k}$ where $(i_1,\dots,i_q) \in \prod_{k=1}^{q} \Sigma_{N_k}$.

First we shall show that T is well-defined. For $(x_1,\ldots,x_q),(y_1,\ldots,y_q)\in\prod_{k=1}^q I_{k,i_k}$, let us note that

$$\begin{aligned}
[T_f g]_\sigma &= \max_{(i_1,\ldots,i_q)\in\prod_{k=1}^q \Sigma_{N_k}} \sup_{(x_1,\ldots,x_q)\neq(y_1,\ldots,y_q)} \frac{|T_f g(x_1,\ldots,x_q)-T_f g(y_1,\ldots,y_q)|}{d_{I^q}((x_1,\ldots,x_q),(y_1,\ldots,y_q))^\sigma}\\
&\le \max_{(i_1,\ldots,i_q)\in\prod_{k=1}^q \Sigma_{N_k}} \Bigg[\sup_{(x_1,\ldots,x_q)\neq(y_1,\ldots,y_q)} \Bigg\{\frac{|f(x_1,\ldots,x_q)-f(y_1,\ldots,y_q)|}{d_{I^q}((x_1,\ldots,x_q),(y_1,\ldots,y_q))^\sigma}\\
&\quad+\frac{|\alpha|\,\big|(g-s)(Q_{i_1\ldots i_q}(x_1,\ldots,x_q))-(g-s)(Q_{i_1\ldots i_q}(y_1,\ldots,y_q))\big|}{d_{I^q}((x_1,\ldots,x_q),(y_1,\ldots,y_q))^\sigma}\Bigg\}\Bigg]\\
&\le [f]_\sigma + N^\sigma|\alpha|\left([g]_\sigma+[s]_\sigma\right).
\end{aligned}$$

For $g,h\in\mathcal{H}_f^\sigma(I^q)$

$$\begin{aligned}
\|T_f g - T_f h\| &= \|T_f g - T_f h\|_\infty + [T_f g - T_f h]_\sigma\\
&\le |\alpha|\|g-h\|_\infty + N^\sigma|\alpha|[g-h]_\sigma\\
&\le N^\sigma|\alpha|\|g-h\|.
\end{aligned}$$

Since $N^\sigma|\alpha|<1$, we have that T is a contraction map on $\mathcal{H}_f^\sigma(I^q)$. Using the Banach contraction principle, we get a unique fixed point of T, namely $f^\alpha_{\Delta,s}\in\mathcal{H}_f^\sigma(I^q)$. □

Theorem 1.10 *Let f and s be such that*

$$\begin{aligned}
|f((x_1,\ldots,x_q))-f((y_1,\ldots,y_q))| &\le K_f d_{I^q}((x_1,\ldots,x_q),(y_1,\ldots,y_q))^\sigma,\\
|s((x_1,\ldots,x_q))-s((x_1,\ldots,x_q))| &\le K_s d_{I^q}((x_1,\ldots,x_q),(y_1,\ldots,y_q))^\sigma.
\end{aligned}\tag{1.3}$$

for every $(x_1,\ldots,x_q),(y_1,\ldots,y_q)\in I^q$, and for fixed $K_f,K_s>0$. Assume that for some $k_f>0,\delta_0>0$ the following holds: for each $(x_1,\ldots,x_q)\in I^q$ and $0<\delta<\delta_0$ there exists $(y_1,\ldots,y_q)$ such that $d_{I^q}((x_1,\ldots,x_q),(y_1,\ldots,y_q))\le\delta$ and

$$|f((x_1,\ldots,x_q))-f((y_1,\ldots,y_q))| \ge k_f d_{I^q}((x_1,\ldots,x_q),(y_1,\ldots,y_q))^\sigma.\tag{1.4}$$

Furthermore, we suppose $N_1=\cdots=N_q=N$, $x_{k,i_k}-x_{k,i_k-1}=\frac{1}{N}$, for all $i_k\in\Sigma_{N_k}$, $k\in\Sigma_q$ and constant scaling function α.
If $|\alpha|<\min\left\{\frac{1}{N},\frac{k_f}{(K_{f^\alpha_{\Delta,s}}+K_s)N^\sigma}\right\}$, then $\dim_B\left(Gr(f^\alpha_{\Delta,s})\right)=q+1-\sigma$.

Proof Proof follows on similar lines of Theorem 5.16 in [29], however, we include the proof for completeness and record. Since $N^\sigma|\alpha|<N|\alpha|<1$, Theorem 1.9 implies that $f^\alpha_{\Delta,s}$ is Hölder continuous with exponent σ. Therefore, $\underline{\dim}_B\big(Gr(f^\alpha_{\Delta,s})\big)\le\overline{\dim}_B\big(Gr(f^\alpha)\big)\le q+1-\sigma$, see, for instance, [17, Corollary 11.2]. It remains to obtain the lower bound. For this, we proceed by recalling the self-referential equation 1.3

$$f^\alpha_{\Delta,s}(x_1,\ldots,x_q)=f(x_1,\ldots,x_q)+\alpha\big[f^\alpha_{\Delta,s}\big(Q_{i_1\ldots i_q}(x_1,\ldots,x_q)\big)-s\big(Q_{i_1\ldots i_q}(x_1,\ldots,x_q)\big)\big],$$

for every $(x_1,\ldots,x_q)\in\prod_{k=1}^q I_{k,i_k}$ and $(i_1,\ldots,i_q)\in\prod_{k=1}^q\Sigma_{N_k}$.
For $(x_1,\ldots,x_q),(y_1,\ldots,y_q)\in\prod_{k=1}^q I_{k,i_k}$ we obtain

$$
\begin{aligned}
&|f^{\alpha}_{\Delta,s}(x_1,\dots,x_q) - f^{\alpha}_{\Delta,s}(y_1,\dots,y_q)| \\
&= \Big| f(x_1,\dots,x_q) - f(y_1,\dots,y_q) + \alpha f^{\alpha}_{\Delta,s}\big(Q_{i_1\dots i_q}(x_1,\dots,x_q)\big) - \alpha f^{\alpha}_{\Delta,s}\big(Q_{i_1\dots i_q}(y_1,\dots,y_q)\big) \\
&\qquad - \alpha s\big(Q_{i_1\dots i_q}(x_1,\dots,x_q)\big) + \alpha s\big(Q_{i_1\dots i_q}(y_1,\dots,y_q)\big)\Big| \\
&\geq |f(x_1,\dots,x_q) - f(y_1,\dots,y_q)| - |\alpha| \Big| f^{\alpha}_{\Delta,s}\big(Q_{i_1\dots i_q}(x_1,\dots,x_q)\big) - f^{\alpha}_{\Delta,s}\big(Q_{i_1\dots i_q}(y_1,\dots,y_q)\big)\Big| \\
&\qquad - |\alpha| \Big| s\big(Q_{i_1\dots i_q}(x_1,\dots,x_q)\big) - s\big(Q_{i_1\dots i_q}(y_1,\dots,y_q)\big)\Big|.
\end{aligned}
$$

Using Equation (1.4), we have

$$
\begin{aligned}
|f^{\alpha}_{\Delta,s}(x_1,\dots,x_q) - f^{\alpha}_{\Delta,s}(y_1,\dots,y_q)| \geq\ & k_f\, d_{I^q}((x_1,\dots,x_q),(y_1,\dots,y_q))^{\sigma} \\
& - |\alpha|\, K_{f^{\alpha}_{\Delta,s}}\, d_{I^q}\big(Q_{i_1\dots i_q}(x_1,\dots,x_q), Q_{i_1\dots i_q}(y_1,\dots,y_q)\big)^{\sigma} \\
& - |\alpha|\, K_s d_{I^q}\big(Q_{i_1\dots i_q}(x_1,\dots,x_q), Q_{i_1\dots i_q}(y_1,\dots,y_q)\big)^{\sigma} \\
\geq\ & k_f\, d_{I^q}((x_1,\dots,x_q),(y_1,\dots,y_q))^{\sigma} \\
& - |\alpha|\, K_{f^{\alpha}_{\Delta,s}} N^{\sigma} d_{I^q}((x_1,\dots,x_q),(y_1,\dots,y_q))^{\sigma} \\
& - |\alpha|\, K_s N^{\sigma} d_{I^q}((x_1,\dots,x_q),(y_1,\dots,y_q))^{\sigma} \\
=\ & k_f\, d_{I^q}((x_1,\dots,x_q),(y_1,\dots,y_q))^{\sigma} \\
& - |\alpha|\, K_{f^{\alpha}} N^{\sigma}\, d_{I^q}((x_1,\dots,x_q),(y_1,\dots,y_q))^{\sigma} \\
& - |\alpha|\, K_s N^{\sigma}\, d_{I^q}((x_1,\dots,x_q),(y_1,\dots,y_q))^{\sigma} \\
=\ & \Big(k_f - (K_{f^{\alpha}} + K_s)|\alpha| N^{\sigma}\Big) d_{I^q}((x_1,\dots,x_q),(y_1,\dots,y_q))^{\sigma}.
\end{aligned}
$$

Let $K := k_f - (K_{f^{\alpha}} + K_s)|\alpha|N^{\sigma}$. Using the above estimates and the definition of lower box dimension, we deduce

$$
\underline{\dim}_B(Gr(f^{\alpha})) = \underline{\lim}_{\delta\to 0} \frac{\log\Big(N_{\delta}(Gr(f^{\alpha}))\Big)}{-\log(\delta)} \geq q + 1 - \sigma.
$$

and hence the proof. □

Remark 1.3 With the assumptions of the above theorem, one may construct dimension preserving approximants for a given function, see, for instance, [30, Theorem 3.16].

Navascués [23] developed the notion of (univariate) α-fractal function via so-called (univariate) fractal operator. In [28, 29], her collaborators extended some of her results in bivariate setting. On putting $s = B_{m_1,\dots,m_q}(f)$ in Equation 1.3, we have a unique function $f^{\alpha}_{\Delta, B_{m_1,\dots,m_q}} \in C(I^q)$ such that

$$
\begin{aligned}
f^{\alpha}_{\Delta,B_{m_1,\dots,m_q}}(x_1,\dots,x_q) &= f(x_1,\dots,x_q) + \alpha(x_1,\dots,x_q)\, f^{\alpha}_{\Delta,B_{m_1,\dots,m_q}}\big(Q_{i_1\dots i_q}(x_1,\dots,x_q)\big) \\
&\quad -\alpha(x_1,\dots,x_q)\, B_{m_1,\dots,m_q}(f)\big(Q_{i_1\dots i_q}(x_1,\dots,x_q)\big), \qquad (1.5)
\end{aligned}
$$

for $(x_1,\dots,x_q) \in \prod_{k=1}^{q} I_{k,i_k}$, $i_k \in \Sigma_{N_k}$.

Following the work of [28], we define a single-valued fractal operator $\mathcal{F}^{\alpha}_{m_1,\dots,m_q} : C(I^q) \to C(I^q)$ by

$$
\mathcal{F}^{\alpha}_{m_1,\dots,m_q}(f) = f^{\alpha}_{\Delta,B_{m_1,\dots,m_q}}.
$$

In [28], several operator theoretic results for (bivariate) fractal operator are obtained. Following Theorem 3.2 of [28], we have that $\mathcal{F}^{\alpha}_{m_1,\ldots,m_q}$ is a bounded linear operator, see, for instance, [27, Remark 3.8].

Lemma 1.3 ([7], Lemma 1) *Let* $(X, \|.\|)$ *be a Banach space,* $T : X \to X$ *be a linear operator. Suppose there exist constants* $\lambda_1, \lambda_2 \in [0, 1)$ *such that*

$$\|Tx - x\| \le \lambda_1\|x\| + \lambda_2\|Tx\|, \quad \textit{for all } x \in X.$$

Then, T is a topological isomorphism, and

$$\frac{1-\lambda_2}{1+\lambda_1}\|x\| \le \|T^{-1}x\| \le \frac{1+\lambda_2}{1-\lambda_1}\|x\|, \quad \textit{for all } x \in X.$$

Note 1.2 We have the following.

$$B_{m_1,\ldots,m_q}(f)(x_1,\ldots,x_q) = \sum_{i_1=0}^{m_1}\cdots\sum_{i_q=0}^{m_q} f\Big(\frac{i_1}{m_1},\cdots,\frac{i_q}{m_q}\Big)\binom{m_1}{i_1}\cdots\binom{m_q}{i_q}x_1^{i_1}(1-x_1)^{m_1-i_1} \ldots x_q^{i_q}(1-x_q)^{m_q-i_q},$$

Choosing $f = 1$, we have

$$\begin{aligned} B_{m_1,\ldots,m_q}1(x_1,\ldots,x_q) &= \sum_{i_1=0}^{m_1}\cdots\sum_{i_q=0}^{m_q} f\Big(\frac{i_1}{m_1},\ldots,\frac{i_q}{m_q}\Big)\binom{m_1}{i_1}\cdots\binom{m_q}{i_q}x_1^{i_1}(1-x_1)^{m_1-i_1} \ldots x_q^{i_q}(1-x_q)^{m_q-i_q}\\ &= \sum_{i_1=0}^{m_1}\binom{m_1}{i_1}x_1^{i_1}(1-x_1)^{m_1-i_1}\cdots\sum_{i_q=0}^{m_q}\binom{m_q}{i_q}x_q^{i_q}(1-x_q)^{m_q-i_q}\\ &= (x_1+1-x_1)^{m_1}\cdots(x_q+1-x_q)^{m_q}\\ &= 1, \end{aligned}$$

It follows that $\|B_{m_1,\ldots,m_q}\| \ge 1$. Now, for every $f \in C(I^q)$ we get

$$\begin{aligned} |B_{m_1,\ldots,m_q}(f)(x_1,\ldots,x_q)| &\le \|f\|_\infty \sum_{i_1=0}^{m_1}\cdots\sum_{i_q=0}^{m_q} f\Big(\frac{i_1}{m_1},\cdots,\frac{i_q}{m_q}\Big)\binom{m_1}{i_1}\cdots\binom{m_q}{i_q}x_1^{i_1} (1-x_1)^{m_1-i_1}\ldots x_q^{i_q}(1-x_q)^{m_q-i_q}\\ &= \|f\|_\infty, \end{aligned}$$

which produces $\|B_{m_1,\ldots,m_q}\| \le 1$. Therefore, we have $\|B_{m_1,\ldots,m_q}\| = 1$.

Theorem 1.11 *The fractal operator* $\mathcal{F}^{\alpha}_{m_1,m_2,\ldots,m_q} : C(I^q) \to C(I^q)$ *is a topological isomorphism.*

Proof Using equation (1.5) and note 1.2, one gets

$$\begin{aligned} \big\|f - \mathcal{F}^{\alpha}_{m_1,\ldots,m_q}(f)\big\|_\infty &\le \|\alpha\|_\infty\big\|\mathcal{F}^{\alpha}_{m_1,\ldots,m_q}(f) - B_{m_1,\ldots,m_q}f\big\|_\infty\\ &= \|\alpha\|_\infty\big\|\mathcal{F}^{\alpha}_{m_1,\ldots,m_q}(f)\big\|_\infty + \|\alpha\|_\infty\|f\|_\infty. \end{aligned}$$

Since $\|\alpha\|_\infty < 1$, the previous lemma yields that the fractal operator $\mathcal{F}^\alpha_{m_1,\ldots,m_q}$ is a topological isomorphism.□

Remark 1.4 For $q = 2$, the above theorem may strengthen item-4 of [28, Theorem 3.2]. To be precise, item-4 tells that $\mathcal{F}^\alpha_{m_1,\ldots,m_q}$ is a topological isomorphism if $\|\alpha\|_\infty < \left(1+\|I - B_{m_1,\ldots,m_q}\|\right)^{-1}$, which is more restricted than the standing assumption considered in the above theorem, that is, $\|\alpha\|_\infty < 1$.

Theorem 1.12 *Let $f \in C(I^q)$ be such that $f(x_1,\ldots,x_q) \geq 0$, for all $x_1,\ldots,x_q \in I^q$. Then for $\epsilon > 0$, and for $\alpha \in C(I^q)$ satisfying $\|\alpha\|_\infty < 1$, there exists an α-fractal function $g^\alpha_{\Delta,B_{m_1,\ldots,m_q}}$ satisfying*

$$g^\alpha_{\Delta,B_{m_1,\ldots,m_q}}(x_1,\ldots,x_q) \geq 0, \ \textit{for all } x_1,\ldots,x_q \in I^q \ \textit{ and } \|f - g^\alpha_{\Delta,B_{m_1,\ldots,m_q}}\|_\infty < \epsilon.$$

Proof Note that the Bernstein operator $B_{m_1,\ldots,m_q}$ fixes the constant function 1, i.e., $B_{m_1,\ldots,m_q}(1) = 1$, where $1(x_1,\ldots,x_q) = 1$ on I^q. Consider $\alpha \in C(I^q)$ such that $\|\alpha\|_\infty < 1$. From Equation 1.5, we deduce

$$\|g^\alpha_{\Delta,B_{m_1,\ldots,m_q}} - g\|_\infty \leq \|\alpha\|_\infty \|g^\alpha_{\Delta,B_{m_1,\ldots,m_q}} - B_{m_1,\ldots,m_q} g\|_\infty, \text{ for all } g \in C(I^q).$$

Choose $g = 1$, then the above inequality gives

$$\|f^\alpha_{\Delta,B_{m_1,\ldots,m_q}} - 1\|_\infty \leq \|\alpha\|_\infty \|f^\alpha_{\Delta,B_{m_1,\ldots,m_q}} - 1\|_\infty,$$

and this further yields $\|f^\alpha_{\Delta,B_{m_1,\ldots,m_q}} - 1\|_\infty = 0$. Therefore, $f^\alpha_{\Delta,B_{m_1,\ldots,m_q}} = 1$, i.e., $\mathcal{F}^\alpha_{m_1,\ldots,m_q}(1) = 1$.
For $\epsilon > 0$, $\alpha \in C(I^q)$ and $f \in C(I^q)$. Using Theorem 1.2, there exists a function $h^\alpha_{\Delta,B_{m_1,\ldots,m_q}}$ such that

$$\|f - h^\alpha_{\Delta,B_{m_1,\ldots,m_q}}\|_\infty < \frac{\epsilon}{2}, \text{ where } \mathcal{F}^\alpha_{m_1,\ldots,m_q}(h) = h^\alpha_{\Delta,B_{m_1,\ldots,m_q}}.$$

Define $g^\alpha_{\Delta,B_{m_1,\ldots,m_q}}(x_1,\ldots,x_q) = h^\alpha_{\Delta,B_{m_1,\ldots,m_q}}(x_1,\ldots,x_q) + \frac{\epsilon}{2}$ for all $x_1,\ldots,x_q \in I^q$.
Since $\mathcal{F}^\alpha_{m_1,\ldots,m_q}(1) = 1$,

$$\begin{aligned} g^\alpha_{\Delta,B_{m_1,\ldots,m_q}}(x_1,\ldots,x_q) &= h^\alpha_{\Delta,B_{m_1,\ldots,m_q}}(x_1,\ldots,x_q) + \frac{\epsilon}{2}1(x_1,\ldots,x_q) \\ &= h^\alpha_{\Delta,B_{m_1,\ldots,m_q}}(x_1,\ldots,x_q) + \frac{\epsilon}{2}1^\alpha(x_1,\ldots,x_q). \end{aligned}$$

Further, since $\mathcal{F}^\alpha_{m_1,\ldots,m_q}$ is a linear operator

$$g^\alpha_{\Delta,B_{m_1,\ldots,m_q}} = h^\alpha_{\Delta,B_{m_1,\ldots,m_q}} + \frac{\epsilon}{2}1^\alpha = \mathcal{F}^\alpha_{m_1,\ldots,m_q}(h + \frac{\epsilon}{2}1).$$

Moreover,

$$\begin{aligned} g^\alpha_{\Delta,B_{m_1,\ldots,m_q}}(x_1,\ldots,x_q) &= h^\alpha_{\Delta,B_{m_1,\ldots,m_q}}(x_1,\ldots,x_q) + \frac{\epsilon}{2} \\ &= h^\alpha_{\Delta,B_{m_1,\ldots,m_q}}(x_1,\ldots,x_q) + \frac{\epsilon}{2} - f(x_1,\ldots,x_q) + f(x_1,\ldots,x_q) \\ &\geq f(x_1,\ldots,x_q) + \frac{\epsilon}{2} - \|h^\alpha_{\Delta,B_{m_1,\ldots,m_q}} - f\|_\infty \\ &\geq 0. \end{aligned}$$

Further, we get

$$\begin{aligned}\|f - g^{\alpha}_{\Delta,B_{m_1,\dots,m_q}}\|_\infty &\le \|f - h^{\alpha}_{\Delta,B_{m_1,\dots,m_q}}\|_\infty + \|h^{\alpha}_{\Delta,B_{m_1,\dots,m_q}} - g^{\alpha}_{\Delta,B_{m_1,\dots,m_q}}\|_\infty \\ &< \frac{\epsilon}{2} + \frac{\epsilon}{2} \\ &= \epsilon,\end{aligned}$$

completing the proof. □

Now we shall try to explore the fractal approximation with an example. We shall take $q = 2$ to study the approximation. If $q = 2$, then the function is known as bivariate function. We shall take the same example as in [[29], Example 3.20] only the changes in scaling functions in our example. Here we have taken a different set of scaling functions.

In the next example, we shall examine the behavior of a fractal function that corresponds to parameters, i.e., the base function s, scaling function α, net Δ and the function f. With the help of the following graphical depiction, we can see that the fractal version of the original function is completely dependent on the scaling function. The variations in the graphs for different values of parameters are really interesting to see. Therefore, we shall select the various parameters corresponding to the same function.

Also it will set the base of our next section, i.e., 1.3, as corresponding to the given germ function we can observe the different set of α-fractal functions, hence to examine the behaviour of these changes we shall introduce the multi-valued map, thus multi-valued section for our chapter.

Example 1.1 Consider the following function in the domain $[-1, 1] \times [-1, 1]$ as the germ function

$$f(x, y) = -x^2 - y^2,$$

which is shown in the figures. Let us consider the following parameters:

1. A net Δ determined by the partition $\{-1, -0.5, 0, 0.5, 1\}$ of $[-1, 1]$.
2. Base function $s(x, y) = x^2y^2 f(x, y)$ and different scaling functions $\alpha(x, y) = 0.2,\ 0.7,\ \frac{x}{3}+\frac{x}{3}$ respectively, for all $(x, y) \in [-1, 1] \times [-1, 1]$, see Figure 8.1.
3. Base function $s(x, y) = (2 - x^2y^2)f(x, y)$ and different scaling functions $\alpha(x, y) = 0.2,\ 0.7,\ \frac{x}{3} + \frac{x}{3}$ respectively, for all $(x, y) \in [-1, 1] \times [-1, 1]$, see Figure 1.2.

1.3 Some Multi-valued Mappings

As already discussed in the previous section and it has been observed through Example 1.1 that the α-fractal function is depends on parameters, base function s, scaling function α, net Δ and the function f. To observe the collective behaviour of this α-fractal function with the changes in these parameters we shall introduce the a set-valued mapping. For this, first we shall collect some preliminaries things about set-valued mappings which will be useful throughout this section.

Definition 1.5 ([30]). Let $(X, \|.\|_X)$ and $(Y, \|.\|_Y)$ be normed linear spaces. For a multi-valued (set-valued) mapping $T : X \rightrightarrows Y$, the domain of T is defined by $\mathrm{Dom}(T) := \{x \in X : T(x) \neq \emptyset\}$. Then, $T : X \rightrightarrows Y$ is

1. *convex* if

$$\lambda T(x_1) + (1 - \lambda)T(x_2) \subseteq T\big(\lambda x_1 + (1 - \lambda)x_2\big), \text{ for all } x_1, x_2 \in \mathrm{Dom}(T),\ \ \lambda \in [0, 1].$$

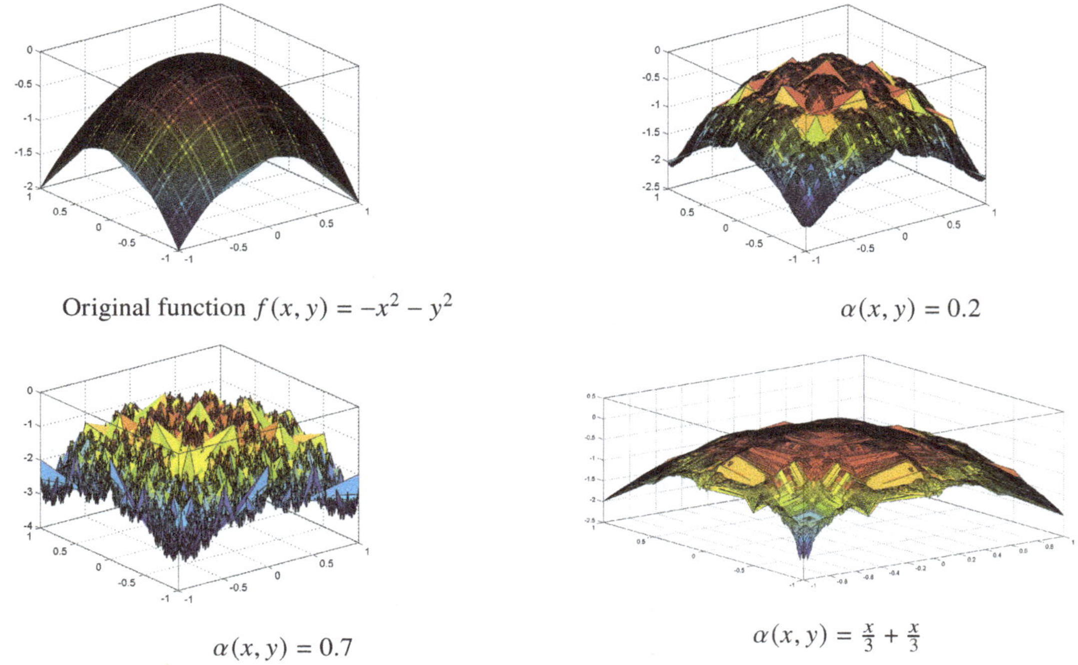
Original function $f(x, y) = -x^2 - y^2$
$\alpha(x, y) = 0.2$
$\alpha(x, y) = 0.7$
$\alpha(x, y) = \frac{x}{3} + \frac{x}{3}$

Fig. 1.1 The bivariate function $f = -x^2 - y^2$ with base function $s(x, y) = x^2 y^2 f(x, y)$ and its fractal perturbations behavior with different choices of scaling functions.

2. *process* if
$$\lambda T(x) = T(\lambda x), \text{ for all } x \in X, \ \lambda > 0, \text{ and } 0 \in T(0).$$
3. *linear* if
$$\beta T(x_1) + \gamma T(x_2) \subseteq T\big(\beta x_1 + \gamma x_2\big), \text{ for all } x_1, x_2 \in \mathrm{Dom}(T), \ \beta, \gamma \in \mathbb{R}.$$
4. *closed* if the graph of T defined by $Gr(T) := \big\{(x, y) \in X \times Y : y \in T(x)\big\}$ is closed.
5. *Lipschitz* if
$$T(x_1) \subseteq T(x_2) + l\|x_1 - x_2\|_X \, U_Y, \text{ for all } x_1, x_2 \in \mathrm{Dom}(T), \text{ for some constant } l > 0,$$
where $U_Y = \{y \in Y : \|y\|_Y \leq 1\}$.
6. *lower semi continuous* at $x \in X$ if there exists a $\Delta > 0$ such that
$$U \cap T(x') \neq \emptyset \text{ whenever } \|x - x'\|_X < \Delta$$
holds for a given open set U in Y satisfying $U \cap T(x) \neq \emptyset$.

Note that the above definitions are also applicable in metric spaces with obvious modifications, see, for instance, [30].

Theorem 1.13 ([10], Corollary 1.4)
Let $T : Dom(T) = X \rightrightarrows Y$ be linear such that $T(0) = \{0\}$. Then, T is single-valued.

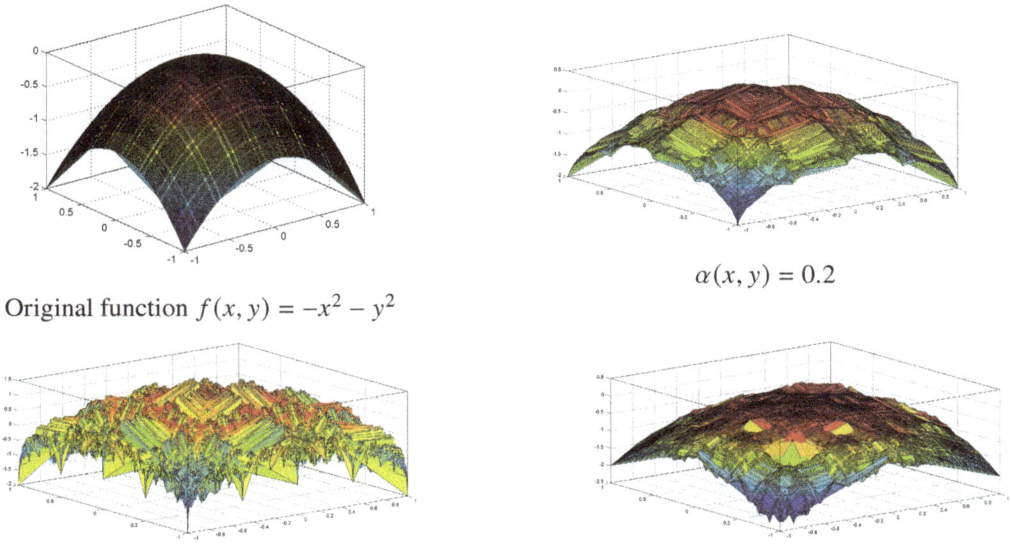

Original function $f(x, y) = -x^2 - y^2$

$\alpha(x, y) = 0.2$

$\alpha(x, y) = 0.7$

$\alpha(x, y) = \frac{x}{3} + \frac{x}{3}$

Fig. 1.2 The bivariate function $f = -x^2 - y^2$ with base function $s(x, y) = (2 - x^2y^2)f(x, y)$ and its fractal perturbations behavior with different choices of scaling functions.

Theorem 1.14 ([10], Corollary 2.1)
Let $T : Dom(T) = X \rightrightarrows Y$ be such that $T(x_0)$ is singleton for some $x_0 \in X$. Then, the following are equivalent:

- *T is single-valued and affine.*
- *T is convex.*

Our work in this part is partly motivated by [24].

Theorem 1.15 *The multi-valued mapping $\mathcal{W}^{\alpha}_{\Delta} : C(I^q) \rightrightarrows C(I^q)$ defined by*

$$\mathcal{W}^{\alpha}_{\Delta}(f) = \{f^{\alpha}_{\Delta, B_{m_1,\ldots,m_q}} : m_1, \ldots, m_q \in \mathbb{N}\}$$

is a Lipschitz process.

Proof Using the linearity of $\mathcal{F}^{\alpha}_{m_1,\ldots,m_q}$, we have

$$\mathcal{W}^{\alpha}_{\Delta}(\lambda f) = \{(\lambda f)^{\alpha}_{\Delta, B_{m_1,\ldots,m_q}} : m_1, \ldots, m_q \in \mathbb{N}\} = \lambda \mathcal{W}^{\alpha}_{\Delta}(f), \text{ for all } f \in C(I^q),\ \lambda > 0.$$

Again by linearity of $\mathcal{F}^{\alpha}_{m_1,\ldots,m_q}$, it is plain that $\mathcal{W}^{\alpha}_{\Delta}(0) = \{0\}$. Therefore, $\mathcal{W}^{\alpha}_{\Delta}$ is a process.

Let $f, g \in C(I^q)$. On applying Equation 1.5, we have

$$\begin{aligned}\left|f^{\alpha}_{\Delta, B_{m_1,\ldots,m_q}}(x_1, \ldots, x_q) - g^{\alpha}_{\Delta, B_{m_1,\ldots,m_q}}(x_1, \ldots, x_q)\right| \leq & \|f - g\|_{\infty} \\ & + \|\alpha\|_{\infty} \|f^{\alpha}_{\Delta, B_{m_1,\ldots,m_q}} - g^{\alpha}_{\Delta, B_{m_1,\ldots,m_q}}\|_{\infty} \\ & + \|\alpha\|_{\infty} \|B_{m_1,\ldots,m_q}(g) - B_{m_1,\ldots,m_q}(f)\|_{\infty},\end{aligned}$$

for any $x_1, \ldots, x_q \in I^q$. Further, we deduce

$$\|f^{\alpha}_{\Delta,B_{m_1,\ldots,m_q}} - g^{\alpha}_{\Delta,B_{m_1,\ldots,m_q}}\|_\infty \leq \frac{1 + \|\alpha\|_\infty \|B_{m_1,\ldots,m_q}\|}{1 - \|\alpha\|_\infty} \|f - g\|_\infty.$$

Using $\|B_{m_1,\ldots,m_q}\| = 1$,

$$\|f^{\alpha}_{\Delta,B_{m_1,\ldots,m_q}} - g^{\alpha}_{\Delta,B_{m_1,\ldots,m_q}}\|_\infty \leq \frac{1 + \|\alpha\|_\infty}{1 - \|\alpha\|_\infty} \|f - g\|_\infty.$$

Consequently, we have

$$\mathcal{W}^{\alpha}_{\Delta}(g) \subseteq \mathcal{W}^{\alpha}_{\Delta}(f) + \frac{1 + \|\alpha\|_\infty}{1 - \|\alpha\|_\infty} \|f - g\|_\infty U_{C(I^q)},$$

proving the Lipschitzness of $\mathcal{W}^{\alpha}_{\Delta}$, and hence the proof. □

Remark 1.5 For the multivalued mapping $\mathcal{W}^{\alpha}_{\Delta}$, let us first note the following:

1. By linearity of $\mathcal{F}^{\alpha}_{\Delta,B_{m_1,\ldots,m_q}}$, we have $\mathcal{W}^{\alpha}_{\Delta}(0) = \{0\}$.
2. Since if $\alpha \neq 0$, $m \neq k$, then $f^{\alpha}_{\Delta,B_{m_1,\ldots,m_q}} \neq f^{\alpha}_{\Delta,B_{k,l}}$, hence $\mathcal{W}^{\alpha}_{\Delta} : C(I^q) \rightrightarrows C(I^q)$ is not single-valued.

In view of the above items, Theorems 1.13–1.14 produce that the mapping $\mathcal{W}^{\alpha}_{\Delta} : C(I^q) \rightrightarrows C(I^q)$ is not convex.

Theorem 1.16 *Let a fixed net* Δ *and* $m_1, \ldots, m_q \in \mathbb{N}$, *the multivalued mapping* $\mathcal{T}^{\Delta}_{m_1,\ldots,m_q} : C(I^q) \rightrightarrows C(I^q)$ *by*

$$\mathcal{T}^{\Delta}_{m_1,\ldots,m_q}(f) = \{f^{\alpha}_{\Delta,B_{m_1,\ldots,m_q}} : \alpha \in C(I^q) \text{ such that } \|\alpha\|_\infty < 1\}$$

is a process.

Proof Let $f \in C(I^q)$ and $\lambda > 0$,

$$\begin{aligned} \lambda\mathcal{T}^{\Delta}_{m_1,\ldots,m_q}(f) =& \lambda\{f^{\alpha} : \alpha \in C(I^q) \text{ such that } \|\alpha\|_\infty < 1\} \\ =& \{\lambda f^{\alpha} : \alpha \in C(I^q) \text{ such that } \|\alpha\|_\infty < 1\} \\ =& \mathcal{T}^{\Delta}_{m_1,\ldots,m_q}(\lambda f). \end{aligned}$$

Furthermore, by applying the linearity of fractal operator, we have $f^{\alpha} = 0$, whenever $f = 0$. That is, $0 \in \mathcal{T}^{\Delta}_{m_1,\ldots,m_q}(0)$. As a result, $\mathcal{T}^{\Delta}_{m_1,\ldots,m_q}$ is a process.

Remark 1.6 One may see that $\mathcal{T}^{\Delta}_{m_1,\ldots,m_q}$ is not convex through the following lines. Let $f, g \in C(I^q)$,

$$\begin{aligned} \mathcal{T}^{\Delta}_{m_1,\ldots,m_q}(f + g) =& \{(f + g)^{\alpha} : \|\alpha\|_\infty < 1\} \\ =& \{f^{\alpha} + g^{\alpha} : \|\alpha\|_\infty < 1\} \\ \subseteq& \{f^{\alpha} + g^{\beta} : \|\alpha\|_\infty < 1, \|\beta\|_\infty < 1\} \\ =& \{f^{\alpha} : \|\alpha\|_\infty < 1\} + \{g^{\beta} : \|\beta\|_\infty < 1\} \\ \subseteq& \mathcal{T}^{\Delta}_{m_1,\ldots,m_q}(f) + \mathcal{T}^{\Delta}_{m_1,\ldots,m_q}(g). \end{aligned}$$

Theorem 1.17 *Let a fixed net* $\triangle$ *and* $m_1, \ldots, m_q \in \mathbb{N}$, *the multivalued mapping* $\mathcal{T}^{\Delta}_{m_1,\ldots,m_q} : C(I^q) \rightrightarrows C(I^q)$ *defined by*

$$\mathcal{T}^{\Delta}_{m_1,\ldots,m_q}(f) = \{f^{\alpha}_{\Delta,B_{m_1,\ldots,m_q}} : \|\alpha\|_\infty \le q < 1\},$$

satisfies the following:

$$\|\mathcal{T}^{\Delta}_{m_1,\ldots,m_q}\| \le 1 + \frac{q}{1-q}\|Id - B_{m_1,\ldots,m_q}\|.$$

Proof We have

$$\begin{aligned}
\|\mathcal{T}^{\Delta}_{m_1,\ldots,m_q}\| &= \sup_{f\in C(I^q)} \frac{d(0,\mathcal{T}^{\Delta}_{m_1,\ldots,m_q}(f))}{\|f\|_\infty} \\
&= \sup_{f\in C(I^q)} \inf_{f^\alpha \in \mathcal{T}^{\Delta}_{m_1,\ldots,m_q}(f)} \frac{\|f^\alpha\|}{\|f\|} \\
&\le \sup_{f\in C(I^q)} \left(1 + \frac{\|\alpha\|_\infty}{1-\|\alpha\|_\infty}\|Id - B_{m_1,\ldots,m_q}\|\right) \\
&\le \sup_{f\in C(I^q)} \left(1 + \frac{q}{1-q}\|Id - B_{m_1,\ldots,m_q}\|\right) \\
&=1 + \frac{q}{1-q}\|Id - B_{m_1,\ldots,m_q}\|,
\end{aligned}$$

hence the proof. □

Theorem 1.18 *For a fixed net* $\triangle$ *and operator* L, *the multivalued mapping* $\mathcal{T}^{\Delta}_{m_1,\ldots,m_q} : C(I^q) \rightrightarrows C(I^q)$ *defined by*

$$\mathcal{T}^{\Delta}_{m_1,\ldots,m_q}(f) = \{f^{\alpha}_{\Delta,B_{m_1,\ldots,m_q}} : \|\alpha\|_\infty < 1\}$$

is lower semicontinuous.

Proof Let $f \in C(I^q)$, let $f^\alpha \in \mathcal{T}^{\Delta}_{m_1,\ldots,m_q}(f)$ and a sequence (f_k) in $C(I^q)$ such that $f_k \to f$. Since the fractal operator is continuous, we have $f^\alpha_k \to f^\alpha$. It is clear that $f^\alpha_k \in \mathcal{T}^{\Delta}_{m_1,\ldots,m_q}(f_k)$. Therefore, the result follows. □

Theorem 1.19 *Let* $\triangle$ *be a net of* I^q *and* $m_1, \ldots, m_q \in \mathbb{N}$. *The multi-valued mapping* $\mathcal{T}^{\Delta}_{m_1,\ldots,m_q} : C(I^q) \rightrightarrows C(I^q)$ *defined by*

$$\mathcal{T}^{\Delta}_{m_1,\ldots,m_q}(f) = \{f^{\alpha}_{\Delta,B_{m_1,\ldots,m_q}} : \|\alpha\|_\infty \le p < 1\},$$

is Lipschitz.

Proof Let $f, g \in C(I^q)$. Equation (1.5) yields

$$\begin{aligned}
\left|f^{\alpha}_{\Delta,B_{m_1,\ldots,m_q}}(x_1,\ldots,x_q) - g^{\alpha}_{\Delta,B_{m_1,\ldots,m_q}}(x_1,\ldots,x_q)\right| &= \|f-g\|_\infty \\
&+ \|\alpha\|_\infty \|f^{\alpha}_{\Delta,B_{m_1,\ldots,m_q}} - g^{\alpha}_{\Delta,B_{m_1,\ldots,m_q}}\|_\infty \\
&+ \|\alpha\|_\infty \|B_{m_1,\ldots,m_q} g - B_{m_1,\ldots,m_q} f\|_\infty,
\end{aligned}$$

for every $x_1, x_2, \ldots, x_q \in I^q$. Further, we deduce

$$\|f^{\alpha}_{\Delta,B_{m_1,\ldots,m_q}} - g^{\alpha}_{\Delta,B_{m_1,\ldots,m_q}}\| \leq \frac{1+\|\alpha\|_\infty\|B_{m_1,\ldots,m_q}\|}{1-\|\alpha\|_\infty}\|f-g\|_\infty.$$

Since $\|\alpha\|_\infty \leq p$ and $\|B_{m_1,\ldots,m_q}\| = 1$, we get

$$\|f^{\alpha}_{\Delta,B_{m_1,\ldots,m_q}} - g^{\alpha}_{\Delta,B_{m_1,\ldots,m_q}}\| \leq \frac{1+p}{1-p}\|f-g\|.$$

Choosing $l = \frac{1+p}{1-p}$, we have

$$\mathcal{T}^{\Delta}_{m_1,\ldots,m_q}(g) \subset \mathcal{T}^{\Delta}_{m_1,\ldots,m_q}(f) + l\,\|f-g\|_\infty U_{C(I^q)},$$

proving the assertion. □

Theorem 1.20 *For a fixed admissible scale vector α and $m_1, \ldots, m_q \in \mathbb{N}$, the multivalued mapping $\mathcal{V}^{\alpha}_{m_1,\ldots,m_q} : C(I^q) \rightrightarrows C(I^q)$ defined by*

$$\mathcal{V}^{\alpha}_{m_1,\ldots,m_q}(f) = \{f^{\alpha}_{\Delta,B_{m_1,\ldots,m_q}} : \textit{all possible net } \Delta\}$$

is a process.

Proof Let $f \in C(I^q)$ and $\lambda > 0$, then

$$\begin{aligned}\lambda\mathcal{V}^{\alpha}_{m_1,\ldots,m_q}(f) =&\lambda\{f^{\alpha}_{\Delta,B_{m_1,\ldots,m_q}} : \text{all possible net } \Delta\}\\ =&\{\lambda f^{\alpha}_{\Delta,B_{m_1,\ldots,m_q}} : \text{all possible net } \Delta\}\\ =&\{(\lambda f)^{\alpha}_{\Delta,B_{m_1,\ldots,m_q}} : \text{all possible net } \Delta\}\\ =&\mathcal{V}^{\alpha}_{m_1,\ldots,m_q}(\lambda f).\end{aligned}$$

The third equality follows from the fact that the fractal operator $\mathcal{F}^{\alpha}_{m_1,\ldots,m_q}$ is a linear operator. Moreover, using linearity of the fractal operator, we have $f^{\alpha}_{\Delta,B_{m_1,\ldots,m_q}} = 0$, whenever $f = 0$. That is, $0 \in \mathcal{V}^{\alpha}_{m_1,\ldots,m_q}(0)$. Therefore, $\mathcal{V}^{\alpha}_{m_1,\ldots,m_q}$ is a process.

Theorem 1.21 *For a fixed admissible scale function α and $m_1, \ldots, m_q \in \mathbb{N}$, the multivalued mapping $\mathcal{V}^{\alpha}_{m_1,\ldots,m_q}$ is lower semicontinuous.*

Proof Let $f \in C(I^q)$, let $f^{\alpha}_{\Delta,B_{m_1,\ldots,m_q}} \in \mathcal{V}^{\alpha}_{m_1,\ldots,m_q}(f)$ and a sequence (f_k) converge to f in $C(I^q)$. Since the fractal operator is continuous, we have $(f_k)^{\alpha}_{\Delta,B_{m_1,\ldots,m_q}} \to f^{\alpha}_{\Delta,B_{m_1,\ldots,m_q}}$. By definition of $\mathcal{V}^{\alpha}_{m_1,\ldots,m_q}$, $(f_k)^{\alpha}_{\Delta,B_{m_1,\ldots,m_q}} \in \mathcal{V}^{\alpha}_{m_1,\ldots,m_q}(f_k)$. Hence, the lower semicontinuity of $\mathcal{V}^{\alpha}_{m_1,\ldots,m_q}$ follows.

Theorem 1.22 *The multi-valued function $\Phi : [\dim(X), \dim(X) + \dim(Y)] \to C(I^q)$ defined by*

$$\Phi(\beta) := \{f \in C(I^q) : \dim(Gr(f)) = \beta\}$$

is lower semicontinuous.

Proof Let U be an open set of $C(I^q)$. In the light of Theorem 1.2, that is, $\Phi(\alpha) = \mathcal{S}_\alpha$ is a dense subset of $C(I^q)$, we obtain

$$\mathcal{S}(\alpha) \cap U \neq \emptyset, \text{ for all } \alpha \in [\dim(X), \dim(X) + \dim(Y)].$$

Now, by the very definition of lower semicontinuous, the result follows. □

Remark 1.7 Note that the multivalued mapping Φ is not closed. To show this, let $f \in C(I^q)$ with $\dim(Gr(f)) > \dim(X)$. Consider a sequence of Lipschitz functions (f_k) converging to f uniformly. It is obvious that $\dim(Gr(f_k)) = \dim(X)$. Now, we have $\big(\dim(X), f_k\big) \to \big(\dim(X), f\big)$ as $n \to \infty$. Using $\big(\dim(X), f_k\big) \in Gr(\Phi)$ and $\big(\dim(X), f_k\big) \to \big(\dim(X), f\big)$ with $\dim(Gr(f)) > \dim(X)$, we get the result.

1.4 Concluding Remarks

The present chapter intended to develop a newly defined notion of constrained approximation termed as dimension preserving approximation for multi-variate functions. The later part has introduced some multi-valued operators associated with multi-variate α-fractal functions. The notion of dimension preserving approximation is new, and demands further developments. In particular, dimension preserving approximation of set-valued mappings may be one of our future investigations.

References

[1] Barnsley, M.F. 1986. Fractal functions and interpolation. Constructive Approximation, 2(1): 303–329.
[2] Navascués, M.A. 2005. Fractal polynomial interpolation. Zeitschrift für Analysis und ihre Anwendungen, 24(2): 401–418.
[3] Navascués, M.A. 2010. Fractal approximation. Complex Analysis and Operator Theory, 4(4): 953–974.
[4] Verma, S. and P.R. Massopust. 2020. Dimension preserving approximation. arXiv preprint arXiv:2002.05061.
[5] Falconer, K. 2004. Fractal Geometry: Mathematical Foundations and Applications. John Wiley & Sons.
[6] Massopust, P.R. 2016. Fractal Functions, Fractal Surfaces, and Wavelets. Academic Press.
[7] Banerjee, S., D. Easwaramoorthy and A. Gowrisankar. 2021. Fractal Functions, Dimensions and Signal Analysis. Springer.
[8] Chandra, S. and S. Abbas. 2021. The calculus of bivariate fractal interpolation surfaces. Fractals, 29(03): 2150066.
[9] Chandra, S. and S. Abbas. 2021. Analysis of mixed weyl-marchaud fractional derivative and box dimensions. Fractals.
[10] Totik, V. 1994. Approximation by Bernstein polynomials. American Journal of Mathematics, 116(4): 995–1018.
[11] Shen, W. 2018. Hausdorff dimension of the graphs of the classical Weierstrass functions. Mathematische Zeitschrift, 289(1): 223–266.
[12] Gal, S.G. 2008. Shape-preserving approximation by real univariate polynomials. Shape-Preserving Approximation by Real and Complex Polynomials, 1–97.
[13] Walter Rudin. 1964. Principles of Mathematical Analysis, 3, McGraw-hill New York.
[14] Mauldin, R.D. and S.C. Williams. 1986. On the Hausdorff dimension of some graphs. Transactions of the American Mathematical Society, 298(2): 793–803.

[15] Falconer, K.J. and J.M. Fraser. 2011. The horizon problem for prevalent surfaces. Mathematical Proceedings of the Cambridge Philosophical Society, 151: 355–372.
[16] DeVore, R. 1968. One-sided approximation of functions. Journal of Approximation Theory, 1(1): 11–25.
[17] Verma, S. and P. Viswanathan. 2020. Parameter identification for a class of bivariate fractal interpolation functions and constrained approximation. Numerical Functional Analysis and Optimization, 41(9): 1109–1148.
[18] Hutchinson, J.E. 1981. Fractals and self-similarity. Indiana University Mathematics Journal, 30(5): 713–747.
[19] Falconer, K.J. 1988. The Hausdorff dimension of self-affine fractals. Mathematical Proceedings of the Cambridge Philosophical Society, 103: 339–350.
[20] Barnsley, M.F., J. Elton, D.P. Hardin and P.R. Massopust. 1989. Hidden variable fractal interpolation functions. SIAM Journal on Mathematical Analysis, 20(5): 1218–1242.
[21] Barnsley, M.F. and P.R. Massopust. 2015. Bilinear fractal interpolation and box dimension. Journal of Approximation Theory, 192: 362–378.
[22] Hardin, D.P. and P.R. Massopust. 1986. The capacity for a class of fractal functions. Communications in Mathematical Physics, 105(3): 455–460.
[23] Hardin, D.P. and P.R. Massopust. 1993. Fractal interpolation functions from $\mathbb{R}^n$ into $\mathbb{R}^m$ and their projections. Zeitschrift für Analysis und ihre Anwendungen, 12(3): 535–548.
[24] Verma, S. and P. Viswanathan. 2020. A fractalization of rational trigonometric functions. Mediterranean Journal of Mathematics, 17: 1–23.
[25] Ruan, H.J. and Q. Xu. 2015. Fractal interpolation surfaces on rectangular grids. Bulletin of the Australian Mathematical Society, 91(3): 435–446.
[26] Verma, S. and P. Viswanathan. 2020. A fractal operator associated with bivariate fractal interpolation functions on rectangular grids. Results in Mathematics, 75(1): 1–26.
[27] Pandey, K.K. and P. Viswanathan. 2021. Multivariate fractal interpolation functions: Some approximation aspects and an associated fractal interpolation operator. arXiv preprint arXiv:2104.02950.
[28] Cazassa, P.G. and O. Christensen. 1997. Perturbation of operators and applications to frame theory. Journal of Fourier Analysis and Applications, 3(5): 543–557.
[29] Deutsch, F. and I. Singer. 1993. On single-valuedness of convex set-valued maps. Set-Valued Analysis, 1(1): 97–103.
[30] Aubin, J.P. and H. Frankowska. 2009. Set-valued Analysis. Springer Science & Business Media.
[31] Agrawal, V., T. Som and S. Verma. 2021. On bivariate fractal approximation. arXiv preprint arXiv:2101.07146, Jan. 2021.

Chapter 2

Fractal Interpolation: From Global to Local, to Nonstationary and Quaternionic

Peter R Massopust

2.1 Introduction

Over the last few decades, fractal interpolation and approximations have been extensively researched. This research originated with [1] where a special set-up was used to define so-called affine fractal interpolation functions. The graphs of these affine fractal interpolation functions are the attractors of a class of iterated function systems and thus geometrically motivated. An analytic construction of general fractal functions originated in [6, 11, 12] where the concept of a Read-Bajractarević operator is first encountered. Numerous constructions of fractal functions based on Read-Bajractarević operators satisfying given interpolation and approximation conditions followed. Some of these constructions are introduced and summarized in [21, 22]. The number of publications in fractal interpolation theory is enormous and the interested reader may want to search for fractal functions using terms such as "hidden variable," "V-variable," "coalescent," "super," and "α-fractal functions" to name just a few.

This chapter intends to introduce the reader to the concept of fractal interpolation and its extensions from a global to a local setting, then to non-stationarity, and finally to a quaternionic setting. In a certain sense, these are the main set-ups with the possible exclusion of *unbounded* fractal interpolation [25]. Understandably such an endeavor must necessarily restrict itself to the main points of each construction and setting. However, the exposition will give the reader an overall perspective of the issues involved and the techniques used and can be used as a starting point for a deeper investigation into each of the topics.

The outline of this chapter is as follows. Section 2 introduces the global setting of fractal interpolation and exhibits a relationship between the Read-Bajractarević operator and the solution to a canonically associated system of functional equations. In the next section, we extend global interpolation to a local setup giving more flexibility to the construction. This type of interpolation is found in deep applications to fractal imaging and fractal compression [5]. In Section 4, the recently introduced concept of non-stationary fractal interpolation is presented and it is shown that backward trajectories allow distinct features to be delineated at different interpolation scales. The final Section 5, describes the novel setting of fractal interpolation in the theory of quaternions and shows that the non-commutative character of quaternions introduces even more intricate fractal patterns.

Centre of Mathematics, Technical University of Munich, Boltzmannstr. 3, 85748 Garching b. Munich, Germany.
Email: massopust@ma.tum.de

2.2 Global Fractal Interpolation

The current section aims to introduce global fractal interpolation and to relate the global fractal interpolant to the solution of a system of functional equations. We see that this system of functional equations defines in a canonical way a Read-Bajractarević (RB) operator and vice versa. This relationship will be encountered several times in the subsequent sections as well.

In the following, $(\mathsf{E}, d_\mathsf{E})$ denotes a normed space and $(\mathsf{F}, d_\mathsf{F})$ a Banach space. For $n \in \mathbb{N}$, we write $\mathbb{N}_n := \{1, \ldots, n\}$ for the initial segment of the natural numbers $\mathbb{N}$ of length n.

For a given normed space $(\mathsf{E}, \|\cdot\|_\mathsf{E})$ and a map $f : \mathsf{E} \to \mathsf{E}$, we define the Lipschitz constant associated with f by

$$\mathrm{Lip}(f) := \sup_{x,y\in\mathsf{E}, x\neq y} \frac{\|f(x) - f(y)\|_\mathsf{E}}{\|x - y\|_\mathsf{E}}.$$

The map f is called Lipschitz if $\mathrm{Lip}(f) < +\infty$ and a contraction (on E) if $\mathrm{Lip}(f) < 1$.

2.2.1 Bounded Solutions

Let X be a nonempty bounded subset of E. Suppose we are given a finite family $\{l_i\}_{i=1}^n$ of injective contractions $\mathsf{X} \to \mathsf{X}$ generating a partition of X in the sense that

$$\forall\, i, j \in \mathbb{N}_n, i \neq j : l_i(\mathsf{X}) \cap l_j(\mathsf{X}) = \emptyset; \tag{2.1}$$

$$\mathsf{X} = \bigcup_{i=1}^{n} l_i(\mathsf{X}). \tag{2.2}$$

For simplicity, we write $\mathsf{X}_i := l_i(\mathsf{X})$.

Given the above set-up, we are looking for a global function $\psi : \mathsf{X} = \bigcup_{i=1}^{n} \mathsf{X}_i \to \mathsf{F}$ satisfying n functional equations of the form

$$\psi(l_i(x)) = q_i(x) + s_i(x)\psi(x), \qquad \text{on } \mathsf{X} \text{ and for } i \in \mathbb{N}_n, \tag{2.3}$$

where for each $i \in \mathbb{N}_n$, s_i is a given bounded function $\mathsf{X} \to \mathbb{R}$ and q_i a bounded function $\mathsf{X} \to \mathsf{F}$. Recall that a function $f : \mathsf{X} \to \mathsf{F}$ is called *bounded* if there exists a finite $M > 0$ such that $\|f(x)\|_\mathsf{F} \leq M$, for all $x \in \mathsf{X}$.

The idea is to consider (2.3) as the fixed point equation for an associated affine operator acting on an appropriately defined function space.

To this end, let $\mathcal{B}(\mathsf{X}, \mathsf{F}) := \{f : \mathsf{X} \to \mathsf{F} : f \text{ is bounded}\}$ denote the the Banach space of bounded functions equipped with the supremums norm $\|f\| := \sup_{x\in\mathsf{X}} \|f(x)\|_\mathsf{F}$.

On the Banach space $\mathcal{B}(\mathsf{X}, \mathsf{F})$, we define an affine operator $T : \mathcal{B}(\mathsf{X}, \mathsf{F}) \to \mathcal{B}(\mathsf{X}, \mathsf{F})$, called a Read-Bajractarević (RB) operator, by

$$Tf(x) = (q_i \circ l_i^{-1})(x) + (s_i \circ l_i^{-1})(x) \cdot (f \circ l_i^{-1})(x), \tag{2.4}$$

for $x \in \mathsf{X}_i$ and $i \in \mathbb{N}_n$, or, equivalently, by

$$\begin{aligned} Tf(x) &= \sum_{i=1}^{n}(q_i \circ l_i^{-1})(x)\,\chi_{\mathsf{X}_i}(x) + \sum_{i=1}^{n}(s_i \circ l_i^{-1})(x)\cdot(f \circ l_i^{-1})(x)\,\chi_{\mathsf{X}_i}(x) \\ &= T(0) + \sum_{i=1}^{n}(s_i \circ l_i^{-1})(x)\cdot(f \circ l_i^{-1})(x)\,\chi_{\mathsf{X}_i}(x), \quad x \in \mathsf{X}, \end{aligned}$$

where χ_S denotes the characteristic function of a set S: $\chi_S(x) = 1$, if $x \in S$, and $\chi_S(x) = 0$, otherwise.

The following result is well-known (see, for instance, [1, 22]) but for the sake of completeness we reproduce the proof. We also refer the interested reader to [31] where a similar set-up is considered.

Theorem 2.1 *The system of functional equations* (2.3) *has a unique bounded solution* $\psi : \mathsf{X} \to \mathsf{F}$ *provided that*

1. *conditions* (2.1) *and* (2.2) *are satisfied, and*
2. $s := \max\limits_{i \in \mathbb{N}_n} \sup\limits_{x \in \mathsf{X}} |s_i(x)| < 1.$

Proof First note that, as the mappings l_i are injective, the right-hand side of (2.3) can be written as the right-hand side of (2.4).

As the functions l_i, q_i, and s_i are all assumed to be bounded, T maps $\mathcal{B}(\mathsf{X}, \mathsf{F})$ into itself. For all $f, g \in \mathcal{B}(\mathsf{X}, \mathsf{F})$, we have that

$$\begin{aligned} \sup_{x \in \mathsf{X}} \|Tf(x) - Tg(x)\|_{\mathsf{F}} &= \max_{i \in \mathbb{N}_n} \sup_{x \in \mathsf{X}_i} \left\|(s_i \circ l_i^{-1})(x)\cdot(f-g)\circ l_i^{-1}(x)\right\|_{\mathsf{F}} \\ &= \max_{i \in \mathbb{N}_n} \sup_{\xi \in \mathsf{X}} \|s_i(\xi)\cdot(f-g)(\xi)\|_{\mathsf{F}} \\ &\leq \max_{i \in \mathbb{N}_n} \sup_{x \in \mathsf{X}} |s_i(x)| \sup_{x \in \mathsf{X}} \|(f-g)(x)\|_{\mathsf{F}}, \end{aligned}$$

from which it follows that

$$\|Tf - Tg\| \leq s\|f - g\|.$$

Hence, T is contractive on the Banach space $\mathcal{B}(\mathsf{X}, \mathsf{F})$ and therefore, by the Banach Fixed Point Theorem, has a unique fixed point $\psi \in \mathcal{B}(\mathsf{X}, \mathsf{F})$. This fixed point solves the functional equations (2.3). □

Remarks

1. The fixed point $\psi \in \mathcal{B}(\mathsf{X}, \mathsf{F})$ of the RB operator T is also called a *bounded fractal function*. In this context, Eq. (2.3) is also referred to as a *self-referential equation* for ψ.
2. The self-referential equation $T\psi = \psi$ expresses the fractal nature of the graph ψ: It is made up of a finite number of copies of itself with each copy being supported on the partitioning sets X_i. Hence, the terminology *fractal function* for ψ.
3. The proof of Banach's Fixed Point Theorem also provides an algorithm for the construction of ψ: Choose *any* function $\psi_0 \in \mathcal{B}(\mathsf{X}, \mathsf{F})$ and iteratively define the following sequence of functions:

$$\psi_k := T\psi_{k-1}, \quad k \in \mathbb{N}.$$

   Then, ψ is given by $\psi = \lim\limits_{k \to \infty} \psi_k$ where the limit is taking with respect to the norm $\|\cdot\|$ on $\mathcal{B}(\mathsf{X}, \mathsf{F})$.

4. The afore-mentioned algorithm for the construction of ψ together with the proof of the Banach Fixed Point Theorem gives an error estimate as well, namely,
$$\|\psi - \psi_k\| \le \frac{s^k}{1-s}\|\psi_1 - \psi_0\|, \quad k \in \mathbb{N}.$$
5. The fixed point ψ depends on n, the partition $(\mathsf{X}_i : i \in \mathbb{N}_n)$, and the functions s_i and q_i with different choices yielding different fractal functions.
6. Emphasizing the dependence of ψ on the functions s_i, the expression s-fractal function can be found in the literature. (See, for instance, [29].) In this context, one considers a fractal function as the image under an operator $\mathcal{F}^s$ associating with a given (non-fractal) function its fractal analogue.
7. Functional equations such as (2.3) exhibit connections to so-called *fractels* [4, 22] and also to the approximation of rough functions [2]. □

Conditions (2.1) and (2.2) cannot be relaxed without adding some compatibility conditions to guarantee that the RB operator T has the form given by Eq. (2.4). Should Eq. (2.1) not be satisfied, one would have to impose in our current setting the following compatibility conditions:

$$\begin{aligned}&\forall x_1, x_2 \in \mathsf{X}:\\ &l_i(x_1) = l_j(x_2) \implies q_i(x_1) + s_i(x_1)\psi(x_1) = q_j(x_2) + s_j(x_2)\psi(x_2).\end{aligned} \tag{2.5}$$

We refer to [31, 32] for more details regarding this issue.

As an application of the above approach to obtain solutions to functional equations of the form (2.3) or equivalently finding the unique fixed points of the associated RB operator (2.4), we provide the following example.

Example 2.1 Let $\mathsf{E} := \mathbb{R} =: \mathsf{F}$ together with the Euclidean norm $|\cdot|$. Further. let $\mathsf{X} := [0, 1) \subset \mathsf{E}$. Assume that we are given two injective contractions $l_i : [0,1) \to [0,1)$, $i = 1, 2$, with $l_1(x) := \frac{1}{3}x$ and $l_2(x) := \frac{2}{3}x + \frac{1}{3}$. Hence, $\mathsf{X}_1 = [0, \frac{1}{3})$ and $\mathsf{X}_2 = [\frac{1}{3}, 1)$. Clearly, $\mathsf{X} = \mathsf{X}_1 \cup \mathsf{X}_2$ and $\mathsf{X}_1 \cap \mathsf{X}_2 = \emptyset$.

Now choose $q_1(x) = -1$, $q_2(x) = x$, $s_1(x) = \frac{1}{2}\sin(x)$, and $s_2(x) := -\frac{2}{3}\cos(x)$. The system of functional equations and the associated RB operator read then

$$\psi(\tfrac{1}{3}x) = -1 + \tfrac{1}{2}\sin(x)\psi(x) \quad \text{and} \quad \psi(\tfrac{2}{3}x + \tfrac{1}{3}) = x - \tfrac{2}{3}\cos(x)\psi(x),$$

and

$$Tf(x) = \begin{cases} -1 + \frac{1}{2}\sin(3x)f(3x), & 0 \le x < \frac{1}{3};\\ 3x - 1 - \frac{2}{3}\cos(\frac{1}{2}(3x-1))f(\frac{1}{2}(3x-1)), & \frac{1}{3} \le x < 1,\end{cases}$$

respectively.

As $s = \frac{2}{3} < 1$, T is contractive. A plot of the solution, respectively, fixed point ψ, is shown in Figure 2.1.

2.2.2 L^p Solutions

In the following section, we set $\mathsf{X} \subset \mathsf{E} := \mathbb{R}^m$ and $\mathsf{Y} := \mathbb{R}^k$ where the Euclidean spaces $\mathbb{R}^m$ and $\mathbb{R}^k$ are endowed with the corresponding canonical Euclidean norms.

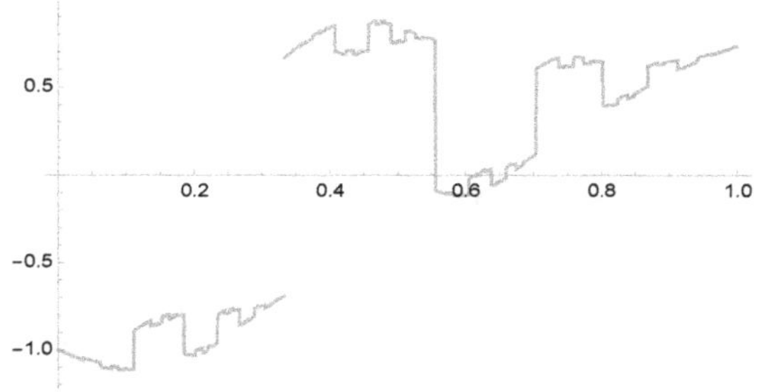

Fig. 2.1 The solution/fixed point ψ.

Recall that the (real) Lebesgue spaces $L^p(\mathsf{X}, \mathbb{R}^k)$, where $\mathsf{X} \subset \mathbb{R}^m$ is nonempty, and are defined as consisting of (equivalence classes of) functions $f : \mathsf{X} \to \mathbb{R}^k$ for which

$$\|f\|_p := \begin{cases} \left(\int_{\mathsf{X}} \|f(x)\|^p dx \right)^{1/p}, & 1 \le p < \infty; \\ \operatorname{ess\,sup}_{x \in \mathsf{X}} \|f(x)\|, & p = \infty. \end{cases}$$

is finite. Here, $\|f(x)\| := \sqrt{|f_1(x)|^2 + \cdots + |f_k(x)|^2}$ with $f := (f_1, \ldots, f_k)$.

We ask under what conditions on the functions q_i and s_i the solution ψ is an element of $L^p(\mathsf{X}, \mathbb{R}^k)$, for $1 \le p < \infty$ and a bounded nonempty set $\mathsf{X} \subset \mathbb{R}^m$.

To this end, note that in order for ψ to be in $L^p(\mathsf{X}, \mathbb{R}^k)$, the RB operator T must map $L^p(\mathsf{X}, \mathbb{R}^k)$ into itself. Therefore, the functions q_i and s_i must also be in $L^p(\mathsf{X}, \mathbb{R}^k)$. Moreover, s_i needs to be in $L^\infty(\mathsf{X}, \mathbb{R}^k)$ for the product $s_i \cdot f$ to be in $L^p(\mathsf{X}, \mathbb{R}^k)$. Thus, as X is bounded it has finite measure and therefore $s_i \in L^\infty(\mathsf{X}, \mathbb{R}^k)$ implies that $s_i \in L^p(\mathsf{X}, \mathbb{R}^k)$ for all $1 \le p \le \infty$.

Now it remains to be shown that the RB operator T is contractive on $L^p(\mathsf{X}, \mathbb{R}^k)$. For this purpose, let $f, g \in L^p(\mathsf{X}, \mathbb{R}^k)$. Then, with $\mathsf{X}_i := l_i(\mathsf{X})$,

$$\begin{aligned} \|Tf - Tg\|_p^p &= \int_{\mathsf{X}} \|Tf(x) - Tg(x)\|^p dx \\ &= \int_{\mathsf{X}} \left\| \sum_{i=1}^n s_i(l_i^{-1}(x)) \cdot (f - g)(l_i^{-1}(x)) \chi_{\mathsf{X}_i}(x) \right\|^p dx \\ &\le \sum_{i=1}^n \int_{\mathsf{X}_i} \left| s_i(l_i^{-1}(x)) \right|^p \left\| (f - g)(l_i^{-1}(x)) \right\|^p dx \\ &= \sum_{i=1}^n \int_{\mathsf{X}} \left| (l_i^{-1})'(x) \right| |s_i(x)|^p \, \|(f - g)((x)\|^p dx. \end{aligned}$$

If, for all $i \in \mathbb{N}_n$, $\left\|(l_i^{-1})'\right\|_\infty =: \lambda_i < \infty$ and $\|s_i\|_\infty =: s_i < \infty$, then

$$\|Tf - Tg\|_p^p \le \left(\sum_{i=1}^n \lambda_i s_i^p \right) \|(f - g)((x)\|_p^p,$$

and T is contractive provided that

$$\sum_{i=1}^{n} \lambda_i s_i^p < 1.$$

Hence, we arrived at the following result

Theorem 2.2 *The system of functional equations*

$$\psi(l_i(x)) = q_i(x) + s_i(x)\psi(x), \quad \text{on } \mathsf{X} \subset \mathbb{R}^m \text{ and for } i \in \mathbb{N}_n,$$

has a unique solution $\psi \in L^p(\mathsf{X}, \mathbb{R}^k)$, $1 \le p \le \infty$, *respectively, the RB operator*

$$Tf(x) = (q_i \circ l_i^{-1})(x) + (s_i \circ l_i^{-1})(x) \cdot (f \circ l_i^{-1})(x), \quad x \in \mathsf{X}_i,\ i \in \mathbb{N}_n,$$

a unique fixed point $\psi \in L^p(\mathsf{X}, \mathbb{R}^k)$ *provided that*

1. $q_i \in L^p(\mathsf{X}, \mathbb{R}^k)$, $s_i \in L^\infty(\mathsf{X}, \mathbb{R}^k)$ *and*
2. $\sum\limits_{i=1}^{n} \lambda_i s_i^p < 1$, *where* $\lambda_i = \left\|(l_i^{-1})'\right\|_\infty$ *and* $s_i = \|s_i\|_\infty$.

Remark 2.1 In a similar fashion, one can derive conditions such that the unique solutions/fixed points ψ are elements of Hölder or Sobolev spaces. See, for instance, [20–22].

2.2.3 Continuous Solutions

So far, we only considered bounded solution/fixed points ψ for (2.3) and (2.4). However, in some instances, a continuous or even differentiable solution is required. We only present a result for continuous ψ and make some remarks about how to obtain differentiable solutions.

Theorem 2.3 *The system of functional equations* (2.3) *has a unique continuous solution* $\psi : \mathsf{X} \to \mathsf{F}$ *provided that*

1. $\mathsf{X} = \bigcup\limits_{i=1}^{n} l_i(\mathsf{X})$,
2. *the functions* l_i, q_i, *and* s_i *are continuous,*
3. *and for all* $i, j \in \mathbb{N}_n$ *and* $x_1, x_2 \in X$:

$$\lim_{x \to x_1} f_j(x) = f_i(x_2) \implies \lim_{x \to x_1} q_j(x) + s_j(x)\psi(x) = q_i(x_2) + s_i(x_2)\psi(x_2). \tag{2.6}$$

Proof We refer the interested reader to [31] or [21]. In the former reference, the proof follows the functional equation setting and in the latter the RB operator setting. □

Example 2.2

We connect up with the previous Example 1 but choose as $\mathsf{X} := [0, 1]$. We modify the functions q_i to be $q_1(x) := x$ and $q_2(x) := 1 - x$. but keep s_1 and s_2 unchanged. Note that here we have $\mathsf{X}_1 \cap \mathsf{X}_2 = \{\frac{1}{3}\}$ and we need to ensure that conditions (2.5) are satisfied. In particular, as we have $l_1(1) = \frac{1}{3} = l_2(0)$, the following equality has to hold:

$$q_1(1) + s_1(1)\psi(1) = q_2(0) + s_2(0)\psi(0). \tag{2.7}$$

The functional equations (2.3) imply for $x \in \{0, 1\}$

$$\psi(0) = q_1(0) + s_1(0)\psi(0) \quad \text{and} \quad \psi(1) = q_2(1) + s_2(1)\psi(0),$$

which gives the values of ψ at the endpoints of X:

$$\psi(0) = \frac{q_1(0)}{1 - s_1(0)} \quad \text{and} \quad \psi(1) = \frac{q_2(1)}{1 - s_2(1)}.$$

The validity of (2.7) guarantees the existence of a *bounded* solution ψ (since $s = \frac{2}{3} < 1$). As $q_1(0) = 0 = q_2(1)$, the solution ψ also vanishes on the boundary of X: $\psi(0) = 0 = \psi(1)$.

In order to obtain a *continuous* solution, equations (2.15) must be satisfied. In our current setting, as all the functions involved are continuous on X, we obtain

$$\lim_{x \to 0-} q_2(x) + s_2(x)\psi(x) = q_1(1) + s_1(1)\psi(1),$$

which is identical to (2.7).

The solution/fixed point ψ is therefore continuous. The graph of ψ is depicted in Figure 2.2.

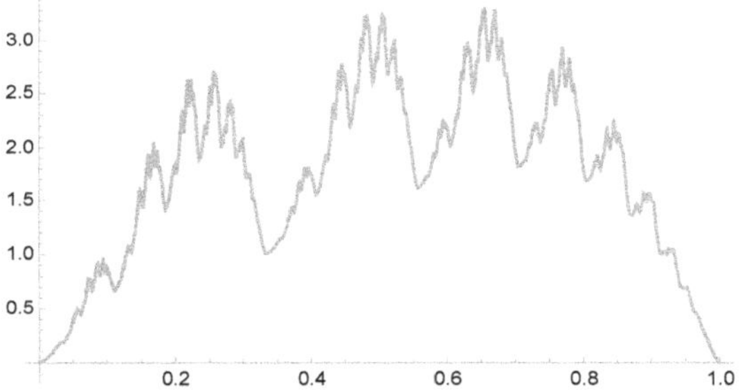

Fig. 2.2 A continuous solution/fixed point ψ.

In [31], a similar setting is considered.

Remark 2.2 For many applications, in particular the setting where $\mathsf{E} := \mathbb{R}$, the partition $(\mathsf{X}_i : i \in \mathbb{N}_n)$ is induced by a given finite set of data points $\{(x_i, y_i) \in \mathbb{R} \times \mathsf{F} : x_0 < x_1 < \cdots < x_n\}$ with $\mathsf{X}_i := [x_{i-1}, x_i]$. The injective contractions l_i are then affine mappings $[x_0, x_n] \to [x_{i-1}, x_i]$. For $\mathsf{F} := \mathbb{R}$, this type of continuous fractal interpolation was first introduced in [1] and the fixed point ψ was termed a *fractal interpolation function* as it was also required that $y_j = \psi(x_j)$, $j \in \{0, 1, \ldots, n\}$.

2.3 Local Fractal Interpolation

In this section, we introduce a generalization of global fractal interpolation. Previously, we considered a fixed nonempty bounded subset X of a normed space E generating a partition of X via a finite family of injective contractions. Now, we replace the single subset X by a finite family of subsets X_i of X.

More precisely, let $\{\mathsf{X}_i : i \in \mathbb{N}_n\}$ be a family of nonempty subsets of a fixed subset X of a normed space E. Suppose $\{l_i\}_{i=1}^n$ is a collection of injective mappings from $\mathsf{X}_i \to \mathsf{X}$ generating a partition of X in the sense that

$$\forall\, i, j \in \mathbb{N}_n, i \neq j : l_i(\mathsf{X}_i) \cap l_j(\mathsf{X}_j) = \emptyset; \tag{2.8}$$

$$\mathsf{X} = \bigcup_{i=1}^{n} l_i(\mathsf{X}_i). \tag{2.9}$$

Note that the l_i need not be contractive mappings here.

Remark 2.3 One can actually have $m < n$ subsets of X and still be able to define n injective mappings satisfying (2.8) and (2.9). In this case some of the injections l_i share the same domain X_j, i.e., X_j is repeated a certain number of times $\mathsf{X}_j = \mathsf{X}_{j+1} = \cdots = \mathsf{X}_{j+r}$. This situation occurs in Section 2.3.3.

Local fractal interpolation looks for local solutions $\psi : \mathsf{X} = \bigcup_{i\in\mathbb{N}_n} l_i(\mathsf{X}_i) \to \mathsf{F}$ of functional equations or for fixed points of RB operators of the form

$$\psi(l_i(x)) = q_i(x) + s_i(x)\psi(x), \quad x \in \mathsf{X}_i,\ i \in \mathbb{N}_n, \tag{2.10}$$

respectively,

$$Tf(x) = (q_i \circ l_i^{-1})(x) + (s_i \circ l_i^{-1})(x) \cdot (f_i \circ l_i^{-1})(x), \quad x \in l_i(\mathsf{X}_i), \quad i \in \mathbb{N}_n, \tag{2.11}$$

where $f_i := f|_{\mathsf{X}_i}$, on appropriate function spaces.

2.3.1 Bounded Local Solutions

The extension of the results presented in Section 2.2 carry over to the setting of local fractal interpolation. However, special care must be taken when considering the domains of the functions involved.

Besides the function space $\mathcal{B}(\mathsf{X}, \mathsf{F})$ already introduced in the previous section, we also need the local version of this space, namely, $\mathcal{B}(\mathsf{X}_i, \mathsf{F})$, $i \in \mathbb{N}_n$. To this end and in view of the result below, we now assume that

1. $s_i \in \mathcal{B}(\mathsf{X}_i, \mathbb{R})$ and
2. $q_i \in \mathcal{B}(\mathsf{X}_i, \mathsf{F})$.

Then the RB operator T defined in (2.11) maps $\mathcal{B}(\mathsf{X}, \mathsf{F})$ into itself. Hence, we arrive at the local version of Theorem 2.1.

Theorem 2.4 *The system of functional equations* (2.10) *has a unique bounded solution* $\psi : \mathsf{X} \to \mathsf{F}$, *respectively, the RB operator* (2.11) *has a unique bounded fixed point* $\psi : \mathsf{X} \to \mathsf{F}$ *provided that*

1. *conditions* (2.8) *and* (2.9) *are satisfied, and*
2. $s := \max_{i\in\mathbb{N}_n} \sup_{x\in\mathsf{X}_i} |s_i(x)| < 1.$

Proof The proof follows the same arguments as those given in the proof of Theorem 2.1; replace X_i and X there by $l_i(\mathsf{X}_i)$ and X_i here. We leave the details to the reader who may also consult [31]. □

As in the global setting of fractal interpolation, a few remarks are in order.

Remarks

1. The solution/fixed point ψ is referred to as a *bounded local fractal function.*
2. The function ψ depends on n, the family of subsets $\{\mathsf{X}_i : i \in \mathbb{N}_n\}$, the partition $(l_i(\mathsf{X}) : i \in \mathbb{N}_n)$ induced by the injective mappings l_i, and the now locally defined functions s_i and q_i. □

As observed above, conditions (2.8) and (2.9) cannot be relaxed without adding some compatibility conditions to guarantee that the RB operator T has the form given by Eq. (2.11). Should Eq. (2.8) not be satisfied, one would have to impose in our current setting the following, now local, compatibility conditions:

$$
\begin{aligned}
&\forall i, j \in \mathbb{N}_n\, \forall x_1 \in \mathsf{X}_i\, \forall x_2 \in \mathsf{X}_j : \\
&\qquad l_i(x_1) = l_j(x_2) \implies q_i(x_1) + s_i(x_1)\psi(x_1) = q_j(x_2) + s_j(x_2)\psi(x_2).
\end{aligned}
\tag{2.12}
$$

We again refer to [31, 32] for more details regarding this issue.

2.3.2 L^p Local Solutions

As an application of local fractal functions for a particular setting, we again consider local fractal functions in L^p spaces for $p \in [0, \infty]$.

For this purpose, we choose the following set up. Let $\mathsf{E} := \mathbb{R} := \mathsf{Y}$ with the canonical Euclidean norm and let $\mathsf{X} := [0, 1]$. Suppose that we are given a partition of X of the form $\Delta := (0 =: x_0 < x_1 < \cdots < x_{n-1} < x_n := 1)$, for some integer $n > 1$. Furthermore, suppose that $\{\mathsf{X}_i : i \in \mathbb{N}_n\}$ is a family of half-open intervals of $[0, 1]$.

We define affine mappings $l_i : \mathsf{X}_i$ onto $[x_{i-1}, x_i)$, $i = 1, \ldots, n-1$, and from $\mathsf{X}_n^+ := \mathsf{X}_n \cup l_n^{-1}(1-)$ onto $[x_{n-1}, x_n]$, where l_n maps X_n onto $[x_{n-1}, x_n)$.

We have the following result for RB-operators defined on the Lebesgue spaces $L^p[0, 1]$, $1 \le p \le \infty$. (See, also, [3].)

Theorem 2.5 *Assume that $q_i \in L^p(\mathsf{X}_i, [0, 1])$ and $s_i \in L^\infty(\mathsf{X}_i, \mathbb{R})$, $i \in \mathbb{N}_n$. The system of functional equations*

$$
\psi(l_i(x)) = q_i(x) + s_i(x)\psi(x), \quad x \in [0, 1],\ i \in \mathbb{N}_n,
$$

has a unique solution $\psi \in L^p[0, 1]$, $1 \le p \le \infty$, respectively, the RB operator

$$
Tf(x) = (q_i \circ l_i^{-1})(x) + (s_i \circ l_i^{-1})(x) \cdot (f_i \circ l_i^{-1})(x), \quad x \in \mathsf{X}_i,\ i \in \mathbb{N}_n,
\tag{2.13}
$$

has a unique fixed point $\psi \in L^p(\Omega, \mathbb{R}^k)$, where $f_i = f|_{\mathsf{X}_i}$, provided that

$$
\begin{cases}
\left(\displaystyle\sum_{i=1}^{n} a_i\, \|s_i\|_{\infty, \mathsf{X}_i}^p \right)^{1/p} < 1, & p \in [1, \infty); \\
\displaystyle\max_{i \in \mathbb{N}_n} \|s_i\|_{\infty, \mathsf{X}_i} < 1, & p = \infty,
\end{cases}
\tag{2.14}
$$

where a_i denotes the Lipschitz constant of $(l_i^{-1})'$. Here, we wrote $\|s_i\|_{\infty, \mathsf{X}_i}$ for $\sup\limits_{x \in \mathsf{X}_i} |s_i(x)|$.

Proof Note that under the hypotheses on the functions q_i and s_i as well as the mappings l_i, Tf is well-defined and an element of $L^p[0, 1]$. It remains to be shown that under condition (2.14), T is contractive on $L^p[0, 1]$.

To this end, let $g, h \in L^p[0,1]$ and let $p \in [0,\infty)$. Then

$$\begin{aligned}
\|Tg - Th\|_p^p &= \int\limits_{[0,1]} |Tg(x) - Th(x)|^p dx \\
&= \int\limits_{[0,1]} \left|\sum_{i=1}^{n} (s_i \circ l_i^{-1})(x)[(g_i \circ l_i^{-1})(x) - (h_i \circ l_i^{-1})(x)]\, \chi_{l_i(\mathbb{X}_i)}(x)\right|^p dx \\
&= \sum_{i=1}^{n} \int\limits_{[x_{i-1},x_i]} \left|(s_i \circ l_i^{-1})(x)[(g_i \circ l_i^{-1})(x) - (h_i \circ l_i^{-1})(x)]\right|^p dx \\
&= \sum_{i=1}^{n} a_i \int\limits_{\mathbb{X}_i} |s_i(x)[g_i(x) - h_i(x)]|^p\, dx \\
&\le \sum_{i=1}^{n} a_i\, \|s_i\|_{\infty,\mathbb{X}_i}^p \int\limits_{\mathbb{X}_i} |g_i(x) - h_i(x)|^p\, dx = \sum_{i=1}^{n} a_i\, \|s_i\|_{\infty,\mathbb{X}_i}^p\, \|f_i - g_i\|_{p,\mathbb{X}_i}^p \\
&= \sum_{i=1}^{n} a_i\, \|s_i\|_{\infty,\mathbb{X}_i}^p\, \|g_i - h_i\|_p^p \le \left(\sum_{i=1}^{n} a_i\, \|s_i\|_{\infty,\mathbb{X}_i}^p\right) \|g - h\|_p^p.
\end{aligned}$$

Now let $p = \infty$. Then,

$$\begin{aligned}
\|Tg - Th\|_\infty &= \left\|\sum_{i=1}^{n} (s_i \circ l_i^{-1})(x)[(g_i \circ l_i^{-1})(x) - (h_i \circ l_i^{-1})(x)]\, \chi_{l_i(\mathbb{X}_i)}(x)\right\|_\infty \\
&\le \max_{i\in\mathbb{N}_n} \left\|(s_i \circ l_i^{-1})(x)[(g_i \circ l_i^{-1})(x) - (h_i \circ l_i^{-1})(x)]\right\|_{\infty,\mathbb{X}_i} \\
&\le \max_{i\in\mathbb{N}_n} \|s_i\|_{\infty,\mathbb{X}_i}\, \|g_i - h_i]\|_{\infty,\mathbb{X}_i} = \max_{i\in\mathbb{N}_n} \|s_i\|_{\infty,\mathbb{X}_i}\, \|g_i - h_i]\|_\infty \\
&\le \left(\max_{i\in\mathbb{N}_n} \|s_i\|_{\infty,\mathbb{X}_i}\right) \|g - h]\|_\infty
\end{aligned}$$

These calculations prove the claims. □

Remark 2.4 Although the conditions (2.14) resemble those presented in Theorem 2.2, these are – because of the local nature of the estimates – more subtle than the former ones.

2.3.3 Continuous Local Solutions

As in the global case, we can consider continuous local solutions to the given set of functional equations, respectively, an RB operator. Here, we present the analog to Theorem 2.3 but refer to the literature for the proof. A good reference in the functional equation setting is again [31]. In the RB operator setting, we refer to [23].

Theorem 2.6 *The system of functional equations* (2.10) *has a unique continuous solution* $\psi : \mathsf{X} \to \mathsf{F}$ *provided that*

1. $\mathsf{X} = \bigcup_{i=1}^{n} l_i(\mathsf{X}_i)$,
2. *the functions* l_i, q_i, *and* s_i *are continuous,*
3. *and* $\forall i, j \in \mathbb{N}_n$, $i \neq j$, $\forall x_1 \in \mathsf{X}$, $\forall x_2 \in \mathsf{X}_i$*:*

$$\lim_{\substack{x \to x_1 \\ x \in \mathsf{X}_j}} f_j(x) = f_i(x_2) \implies \lim_{\substack{x \to x_1 \\ x \in \mathsf{X}_j}} q_j(x) + s_j(x)\psi(x) = q_i(x_2) + s_i(x_2)\psi(x_2). \tag{2.15}$$

As an application and an example for a continuous local fractal interpolation, we present the following set up which plays an important role in fractal-based numerical analysis as discussed in [3].

Suppose that $\mathsf{E} := \mathbb{R} =: \mathsf{F}$ and $\mathsf{X} := [0, 1]$. For an even integer $n \in \mathbb{N}$, define subsets

$$\mathsf{X}_{2j-1} := \mathsf{X}_{2j} := \left[\tfrac{2j-2}{n}, \tfrac{2j}{n}\right], \quad j \in \{1, \ldots, \tfrac{n}{2}\} \tag{2.16}$$

and affine mappings $l_i : \mathsf{X}_i \to [0, 1]$ by

$$l_{2j-1}(x) := \tfrac{x}{2} + \tfrac{j-1}{n} \quad \text{and} \quad l_{2j}(x) := \tfrac{x}{2} + \tfrac{j}{n}, \quad x \in \mathsf{X}_{2j-1} = \mathsf{X}_{2j}. \tag{2.17}$$

Note that

$$l_i(\mathsf{X}_i) = \left[\tfrac{i-1}{n}, \tfrac{i}{n}\right], \quad i \in \mathbb{N}_n.$$

Further, let $x_i := l_i(\mathsf{X}_i) \cap l_{i+1}(\mathsf{X}_{i+1}) = \{\frac{i}{n}\}$, $i \in \mathbb{N}_{n-1}$, $x_0 := 0$, and $x_n := 1$. In the terminology of [31], the elements of $\{x_i : i \in \mathbb{N}_{n-1}\}$ are called *contact points.*

We denote the distinct endpoints of the partitioning intervals $\{l_i(\mathbb{X}_i)\}$ by $\{x_0 < x_1 < \ldots < x_N\}$ where $x_0 = 0$ and $x_N = 1$ and refer to them as knots.

Furthermore, we assume that we are given interpolation values at the endpoints of the intervals $\mathbb{X}_{2j-1} = \mathbb{X}_{2j}$:

$$\Delta := \left\{(x_{2j}, y_j) : j = 0, 1, \ldots, n/2\right\}. \tag{2.18}$$

Let

$$C_\Delta(\mathsf{X}) := \{f \in C(\mathsf{X}) : f(x_{2j}) = y_j, \ \forall\, j = 0, 1, \ldots, n/2\}.$$

Here, $C(\mathsf{X}) := C(\mathsf{X}, \mathbb{R})$ denotes the Banach space of all continuous functions $\mathsf{X} \to \mathbb{R}$ endowed with the supremum norm $\|\cdot\|$. Note that $C_\Delta(\mathsf{X})$ is a closed, hence complete, *metric* subspace of $C(\mathsf{X})$ to which we can apply the Banach fixed point theorem. To this end, consider an RB operator T of the form (2.13) acting on $C_\Delta(\mathsf{X})$.

In order for T to map $C_\Delta(\mathsf{X})$ into itself we require that

$$q_i, s_i \in C(\mathsf{X}_i) := C(\mathsf{X}_i, \mathbb{R}) := \{f : \mathsf{X}_i \to \mathbb{R} : f \text{ continuous}\} \tag{2.19}$$

and that

$$y_{j-1} = (Tf)(x_{2(j-1)}) \quad \wedge \quad y_j = (Tf)(x_{2j}), \quad j = 1, \ldots, n/2, \tag{2.20}$$

where $x_{2j} := \frac{2j}{n}$.

We remark that the preimages of the knots $x_{2(j-1)}$ and x_{2j} are the endpoints of $\mathsf{X}_{2j-1} = \mathsf{X}_{2j}$. Substitution of T, as given in (2.13), into (2.20) and simplification results in

$$\begin{aligned} q_{2j-1}(x_{2(j-1)}) + \left(s_{2j-1}(x_{2(j-1)}) - 1\right) y_{j-1} &= 0, \\ q_{2j}(x_{2j}) + \left(s_{2j}(x_{2j}) - 1\right) y_j &= 0. \end{aligned} \tag{2.21}$$

To ensure global continuity of Tf on $\mathsf{X} = [0, 1]$, we also have to impose the following join-up conditions at the oddly indexed knots. (Note that these oddly indexed knots are the images of the midpoints of the intervals $\mathsf{X}_{2j-1} = \mathsf{X}_{2j}$.)

$$(Tf)(x_{2j-1}-) = (Tf)(x_{2j-1}+), \quad j = 1, \ldots, n/2. \tag{2.22}$$

These join-up conditions imply that

$$q_{2j}(x_{2(j-1)}) + s_{2j}(x_{2(j-1)})y_{j-1} = q_{2j-1}(x_{2j}) + s_{2j-1}(x_{2j})y_j. \tag{2.23}$$

In the case that all functions q_i and s_i are constant, (2.23) reduces to the condition given in [3, Example 2].

We summarize these results in the next theorem.

Theorem 2.7 *Let* $\mathsf{X} := [0, 1]$ *and let* $n \in 2\mathbb{N}$. *Suppose that subsets of* X *are given by* (2.16) *and the associated mappings* l_i *by* (2.17). *Further suppose that the functions* q_i *and* s_i *satisfy* (2.19) *and that the join-up conditions* (2.20) *and* (2.22) *hold. Then, an RB operator* T *of the form* (2.13) *maps* $C_\Delta(\mathsf{X})$ *into itself and is well-defined.*

If, in addition,

$$\max\left\{\|s_i\|_{\infty,\mathsf{X}_i} : i \in \mathbb{N}_n\right\} < 1, \tag{2.24}$$

then T *is a contraction on* $C_\Delta(\mathsf{X})$ *and thus possesses a unique continuous fixed point* $\psi : [0, 1] \to \mathbb{R}$ *satisfying* $\psi \in C_\Delta(\mathsf{X})$.

This unique fixed point ψ is called a *continuous local fractal function.*

Proof It remains to show that under the condition (2.24), T is contractive on $C_\Delta(\mathsf{X})$. This, however, follows immediately from the case $p = \infty$ in the proof of Theorem 2.5. □

For RB operators of the form (2.13) mapping function spaces such as Hölder, Sobolev, Besov or Triebel Lizorkin into themselves, we refer the interested reader to [23, 24].

2.4 Non-stationary Fractal Interpolation

In this section, we extend the notion of global fractal interpolation to a non-stationary setting. In other words, we no longer assume that we keep the functions q_i and s_i the same at each level of iteration in the construction of the fixed point ψ.

2.4.1 The Non-stationary Setting

To this end, consider a doubly-indexed family of injective contractions $\{l_{i_k,k} : i_k \in \mathbb{N}_{n_k},\ k \in \mathbb{N}\}$ from $\mathsf{X} \to \mathsf{X}$ where X is nonempty bounded subset of a normed space E generating a partition of X for each $k \in \mathbb{N}$ in the sense of (2.8) and (2.9).

Suppose that F is a Banach space, $\{q_{i_k,k} : i_k \in \mathbb{N}_{n_k},\ k \in \mathbb{N}\} \subset \mathcal{B}(\mathsf{X}, \mathsf{F})$, and $\{s_{i_k,k} : i_k \in \mathbb{N}_{n_k},\ k \in \mathbb{N}\} \subset \mathcal{B}(\mathsf{X}, \mathbb{R})$ are such that

$$s := \sup_{k\in\mathbb{N}} \max_{i_k\in\mathbb{N}_k} \|s_{i_k,k}\|_\infty < 1.$$

For each $k \in \mathbb{N}$, we define an RB operator $T_k : \mathcal{B}(\mathsf{X}, \mathsf{F}) \to \mathcal{B}(\mathsf{X}, \mathsf{F})$ by

$$(T_k f)(l_{i_k,k}(x)) := q_{i_k,k}(x) + s_{i_k,k}(x) \cdot f(x), \quad \forall x \in \mathsf{X}. \tag{2.25}$$

It is not difficult to verify that each T_k is a contraction on $\mathcal{B}(\mathsf{X},\mathsf{F})$ with Lipschitz constant

$$\operatorname{Lip}(T_k) = \max_{i_k \in \mathbb{N}_k} \|s_{i_k,k}\|_\infty \le s < 1. \tag{2.26}$$

In order to continue, we require the following definition and result from [16] adapted to our current setting.

Definition 2.1 [16, Definition 3.6] Let $\{T_k\}_{k\in\mathbb{N}}$ be a sequence of transformations $T_k : \mathcal{B}(\mathsf{X},\mathsf{F}) \to \mathcal{B}(\mathsf{X},\mathsf{F})$. A subset $\mathcal{I}$ of $\mathcal{B}(\mathsf{X},\mathsf{F})$ is called an invariant set of the sequence $\{T_k\}_{k\in\mathbb{N}}$ if

$$\forall\, k \in \mathbb{N}\; \forall\, x \in \mathcal{I} : T_k(x) \in \mathcal{I}.$$

A criterion for obtaining an invariant domain for a sequence $\{T_k\}_{k\in\mathbb{N}}$ of transformations on $\mathcal{B}(\mathsf{X},\mathsf{F})$ is also given in [16].

Proposition 2.1 *[16, Lemma 3.7]Let $\{T_k\}_{k\in\mathbb{N}}$ be a sequence of transformations on $\mathcal{B}(\mathsf{X},\mathsf{F})$. Suppose there exists a $g \in \mathcal{B}(\mathsf{X},\mathsf{F})$ such that for all $f \in \mathcal{B}(\mathsf{X},\mathsf{F})$*

$$\|T_k(x) - g\| \le \mu\, \|f - g\| + M,$$

for some $\mu \in [0,1)$ and $M > 0$. Then the ball $B_r(g)$ of radius $r = M/(1-\mu)$ centered at g is an invariant set for $\{T_k\}_{k\in\mathbb{N}}$.

Proof The proof, although in a more general setting, is found in [16]. □

Proposition 2.2 *Let $\{T_k\}_{k\in\mathbb{N}}$ be a sequence of RB operators of the form* (2.25) *on* $(\mathcal{B}(\mathsf{X},\mathsf{F}), \|\cdot\|)$. *Suppose that the elements of $\{q_{i_k,k} : i_k \in \mathbb{N}_{n_k},\ k \in \mathbb{N}\}$ satisfy*

$$\sup_{k\in\mathbb{N}} \max_{i_k\in\mathbb{N}_k} \left\|q_{i_k,k}\right\| \le M, \tag{2.27}$$

for some $M > 0$. Then the ball $B_r(0)$ of radius $r = M/(1-s)$ centered at $0 \in \mathcal{B}(\mathsf{X},\mathsf{F})$ is an invariant set for $\{T_k\}_{k\in\mathbb{N}}$.

Proof Let $x \in \mathsf{X}$. Then there exists an $i_k \in \mathbb{N}_{n_k}$ with $x \in l_{i_k,k}(\mathsf{X})$. Thus, for any $f \in \mathcal{B}(\mathsf{X},\mathsf{F})$,

$$\|T_k f(x)\|_{\mathsf{F}} \le \left\| s_{i_k,k} \circ l_{i_k,k}^{-1}(x) \cdot f \circ l_{i_k,k}^{-1}(x) \right\|_{\mathsf{F}} + \|T_k(0)\|_{\mathsf{F}}$$

By (2.27), $T_k(0)$ is uniformly bounded on $\mathcal{B}(\mathsf{X},\mathsf{F})$ by $M > 0$. Hence,

$$\left\| s_{i_k,k} \circ l_{i_k,k}^{-1}(x) \cdot f \circ l_{i_k,k}^{-1}(x) \right\|_{\mathsf{F}} \le s \left\| f \circ l_{i_k,k}^{-1}(x) \right\|_{\mathsf{F}},$$

which gives, after taking the supremum over $x \in \mathsf{X}$, $\|T_k f\| \le s\,\|f\| + M$. Proposition (2.1) now yields the statement. □

To arrive at the main result of this section, we require yet another definition. (See also [16, Section 4].)

Definition 2.2 Suppose that $f_0 \in \mathcal{B}(\mathsf{X},\mathsf{F})$ and that $\{T_k\}_{k\in\mathbb{N}}$ be a sequence of RB operators on $\mathcal{B}(\mathsf{X},\mathsf{F})$. The sequences

$$\Phi_k(f_0) := T_k \circ T_{k-1} \circ \cdots \circ T_1(f_0) \tag{2.28}$$

and

$$\Psi_k(f_0) := T_1 \circ T_2 \circ \cdots \circ T_k(f_0) \tag{2.29}$$

are called the forward, respectively, backward trajectory of f_0.

In connection to the above definition, we need the following result from [16, Corollary 11] adapted to our setting.

Theorem 2.8 *Suppose that $\{T_k\}_{k\in\mathbb{N}}$ is a sequence of RB operators of the form* (2.25) *on $\mathcal{B}(\mathsf{X},\mathsf{F})$. Further suppose that*

1. *there exists a nonempty closed invariant set $\mathcal{J} \subseteq \mathcal{B}(\mathsf{X},\mathsf{F})$ for $\{T_k\}_{k\in\mathbb{N}}$;*
2. *and*

$$\sum_{k=1}^{\infty}\prod_{j=1}^{k}\mathrm{Lip}(T_j) < \infty. \tag{2.30}$$

Then the backward trajectories $\Psi_k(f_0)$ converge for any initial $f_0 \in \mathcal{J}$ to a unique function $\psi \in \mathcal{J}$.

With these preliminaries, we obtain the main result of this section.

Theorem 2.9 *The backwards trajectories $\{\Psi_k\}_{k\in\mathbb{N}}$ converge for any initial $f_0 \in \mathcal{J}$ to a unique function $\psi \in \mathcal{J}$, where $\mathcal{J}$ is the closed ball in $\mathcal{B}(\mathsf{X},\mathsf{F})$ of radius $M/(1-s)$ centered at* 0.

Proof By Proposition 2.2 and Theorem 2.8 it remains to be shown that (2.30) is satisfied. This, however, follows directly from (2.26):

$$\prod_{j=1}^{k}\mathrm{Lip}(T_j) \le s^k \quad \text{and} \quad \sum_{k=1}^{\infty} s^k = \frac{s}{1-s} < \infty.$$

A fixed point ψ generated by a sequence $\{T_k\}$ of non-stationary RB operators will be called a *non-stationary fractal function (of class $\mathcal{B}(X,Y)$).*

Employing a non-stationary sequence of RB operators $\{T_k\}$ allows the construction of more general fractal functions exhibiting different local behavior at different scales. This is illustrated by the following example which is taken from [26].

Example 2.3 Let $\mathsf{X} := [0,1] \subset \mathbb{R}$ and $\mathsf{F} := \mathbb{R}$. Consider the two RB operators

$$T_1 f(x) := \begin{cases} 2x + \frac{1}{2}f(2x), & x \in [0,\frac{1}{2}), \\ 2 - 2x + \frac{1}{2}f(2x-1), & x \in [\frac{1}{2},1], \end{cases}$$

and

$$T_2 f(x) := \begin{cases} 2x + \frac{1}{4}f(2x), & x \in [0,\frac{1}{2}), \\ 2 - 2x + \frac{1}{4}f(2x-1), & x \in [\frac{1}{2},1]. \end{cases}$$

For both operators, $l_i(x) := \frac{1}{2}(x+i-1)$, $i = 1,2$.

The sequence $T_1^k f \to \tau$, where τ denotes the Takagi function [33] and $T_2^k \to q$, where $q(x) = 4x(1-x)$.

Consider the alternating sequence $\{T_k\}_{k\in\mathbb{N}}$ of RB operators given by

$$T_k := \begin{cases} T_1, & 10(j-1) < k \le 10j-5, \\ T_2, & 10j-5 < k \le 10j, \end{cases} \quad j \in \mathbb{N}.$$

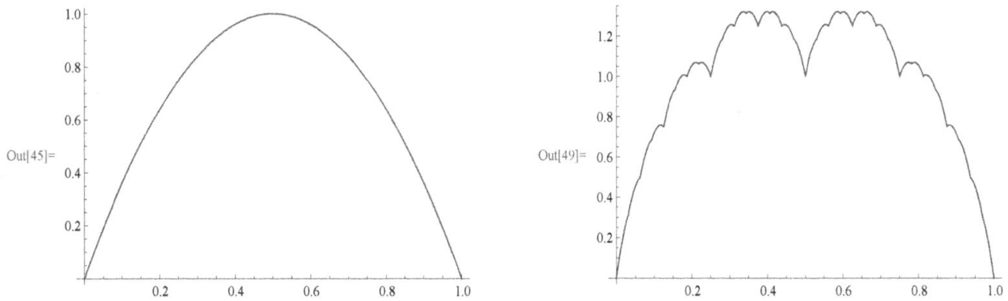

Fig. 2.3 The hybrid $\tau - q$ attractor. It is smooth at one scale but fractal at another.

Two images of this hybrid attractor of the backward trajectory Ψ_k starting with $f_0 \equiv 0$ are shown in Figure 2.3.

2.4.2 Non-stationary Fractal Interpolation

In this subsection, we consider the case $\mathsf{X} := [0, 1]$ and $\mathsf{F} := \mathbb{R}$. In the following, we use a sequence $\{T_k\}$ of RB operators of a particular form to obtain a continuous fixed point ψ. See also [26].

To this end, consider an RB operator defined by (2.4), choose two arbitrary functions $f, b \in \mathcal{B}(\mathsf{X}) := \mathcal{B}(\mathsf{X}, \mathbb{R})$ and set

$$q_i := f \circ l_i - s_i \cdot b. \tag{2.31}$$

The (stationary) RB operator associated with this particular choice is then given by

$$(Tg)(x) = f + s_i(l_i^{-1}(x)) \cdot (g - b)(l_i(x)), \quad x \in l_i(\mathsf{X}) \tag{2.32}$$

Remark 2.5 The fixed point of the above defined RB operator may be thought of as the "fractalization" of a given function f.

To work in a non-stationary setting, let $k \in \mathbb{N}$ and let $\{l_{i_k,k} : i_k \in \mathbb{N}_{n_k},\ k \in \mathbb{N}\}$ be the family of injections from $[0, 1] \to [0, 1]$ generating a partition of $[0, 1]$ in the sense of (2.8) and (2.9). We assume w.l.o.g. that $l_{1,k}(0) = 0$ and $l_{n_k,k}(1) = 1$ and define

$$x_{i_k-1,k} := l_{i_k,k}(0), \quad x_{i_k,k} := l_{i_k,k}(1), \quad i_k \in \mathbb{N}_{n_k}$$

where $x_{0,k} := 0$ and $x_{n_k,k} := 1$. By relabelling – if necessary – we may further assume that $0 = x_{0,k} < \cdots < x_{i_k-1,k} < x_{i_k,k} < \cdots x_{n_k,k} = 1$.

Let $f \in C(\mathsf{X})$ be arbitrary. Define a metric subspace of $C(\mathsf{X})$ by

$$C_*(\mathsf{X}) := \{g \in C(\mathsf{X}) : g(0) = f(0) \wedge g(1) = f(1)\}$$

and note that $C_*(\mathsf{X})$ becomes a complete linear metric space when endowed with the metric d that is induced by the sup-norm from $C(\mathsf{X})$.

Furthermore, let $b \in C_*[0,1]$ be the unique affine function whose graph connects the points $(0, f(0))$ and $(1, f(1))$:

$$b(x) = (f(1) - f(0))x + f(0). \tag{2.33}$$

Let $\{\mathcal{P}_k\}_{k\in\mathbb{N}}$ be a family of sets of points in $\mathsf{X} \times \mathsf{F}$ where

$$\mathcal{P}_k := \{(x_{j_k}, f(x_{j,k}) \in \mathsf{X} \times \mathsf{F} : j = 0, 1, \ldots, n\}.$$

For $k \in \mathbb{N}$, define an RB operator $T_k : C_*[0,1] \to C_*[0,1]$ by

$$T_k g = f + \sum_{i_k=1}^{n_k} s_{i_k,k} \circ l_{i_k,k}^{-1} \cdot (g - b) \circ l_{i_k,k}^{-1} \chi_{l_{i_k,k}[0,1]}, \tag{2.34}$$

where $\{s_{i_k,k}\}_{i_k=1}^{n_k} \subset C[0,1]$ such that

$$\sup_{k\in\mathbb{N}} \max_{i_k \in \mathbb{N}_{i_k}} \|s_{i_k,k}\|_\infty < 1.$$

Note that

$$\begin{aligned} T_k g(x_{i_k,k}-) &= f(x_{i_k,k}-) + s_{i_k,k} \circ l_{i_k,k}^{-1}(x_{i_k,k}-) \cdot (g - b) \circ l_{i_k,k}^{-1}(x_{i_k,k}-) \\ &= f(x_{i_k,k}) + s_{i_k,k}(1) \cdot (f - b)(1) = f(x_{i_k,k}) \end{aligned}$$

and

$$\begin{aligned} T_k g(x_{i_k,k}+) &= f(x_{i_k,k}+) + s_{i_k+1,k} \circ l_{i_k+1,k}^{-1}(x_{i_k,k}+) \cdot (g - b) \circ l_{i_k+1,k}^{-1}(x_{i_k,k}+) \\ &= f(x_{i_k,k}) + s_{i_k+1,k}(0) \cdot (f - b)(0) = f(x_{i_k,k}). \end{aligned}$$

implying that $T_k g$ is continuous at the points $x_{i_k,k} \in [0,1]$:

$$T_k g(x_{i_k,k}-) = T_k g(x_{i_k,k}+), \quad \forall\, i_k \in \{1, \ldots, n-1\}.$$

Hence, $T_k g \in C_*[0,1]$ and $T_k g$ interpolates $\mathcal{P}_k$ in the sense that

$$T_k g(x_{i_k,k}) = f(x_{i_k,k}), \quad \forall\, i_k \in \mathbb{N}_{n_k}.$$

Proposition 2.3 *A nonempty closed invariant set for $\{T_k\}_{k\in\mathbb{N}}$ is given by the closed ball in $C_*(\mathsf{X})$,*

$$\mathcal{I} = \left\{ g \in C_*[0,1] : \|g\| \le \frac{\|f\| + s\|b\|}{1 - s} \right\}, \tag{2.35}$$

where s is defined by (2.26).

Proof From the form (2.31) of the functions $q_{i_k,k}$, we obtain from (2.27) the estimate $\|q_{i_k,k}\| \le \|f\| + s\|b\|$, which by Proposition 2.1 yields the result. □

Together with Theorem 2.9, the above arguments prove the next theorem.

Theorem 2.10 *Let $\{T_k\}_{k\in\mathbb{N}}$ be a sequence of RB operators of the form* (2.34) *each of whose elements acts on the complete metric space $(C_*(\mathsf{X}), d)$ where $f \in C_*(\mathsf{X})$ is arbitrary and b is given by* (2.33)*. Furthermore,* *let the family of functions* $\{s_{i_k,k}\} \subset C(\mathsf{X})$ *satisfy* (2.26). *Then, for any* $f_0 \in \mathcal{J}$, *the backward trajectories* $\Psi_k(f_0)$ *converge to a function* $\psi \in \mathcal{J}$ *which interpolates* $\mathcal{P}_k$.

We refer to the fixed point $\psi \in C_*(\mathsf{X})$ as a *continuous non-stationary fractal interpolation function.*

Remark 2.6 As f_0 one may choose f or b.

To illustrate the above results, we present the following example from [26].

Example 2.4 Consider the two RB operators $T_i : C[0,1] \to C[0,1], i = 1, 2$, defined by

$$(T_1 f)(x) = \begin{cases} -\frac{1}{2} f(4x), & x \in [0, \frac{1}{4}), \\ -\frac{1}{2} + \frac{1}{2} f(4x-1), & x \in [\frac{1}{4}, \frac{1}{2}), \\ \frac{1}{2} f(4x-2), & x \in [\frac{1}{2}, \frac{3}{4}), \\ \frac{1}{2} + \frac{1}{2} f(4x-3), & x \in [\frac{3}{4}, 1], \end{cases}$$

and

$$(T_2 f)(x) := \begin{cases} \frac{3}{4} f(2x), & x \in [0, \frac{1}{2}), \\ \frac{3}{4} + \frac{1}{4} f(2x-1), & x \in [\frac{1}{2}, 1]. \end{cases}$$

The RB operators T_1 and T_2 generate *Kiesswetter's fractal function* [17], respectively, a *Casino function* [10].

Consider again the alternating sequence $\{T_i\}_{i \in \mathbb{N}}$ of RB operators given by

$$T_k := \begin{cases} T_1, & 10(j-1) < k \leq 10j - 5, \\ T_2, & 10j - 5 < k \leq 10j, \end{cases} \qquad j \in \mathbb{N}.$$

Two images of the hybrid attractor of the backward trajectory Ψ_k starting with the function $f_0(x) = x, x \in \mathsf{X}$, are shown below in Figure 2.4.

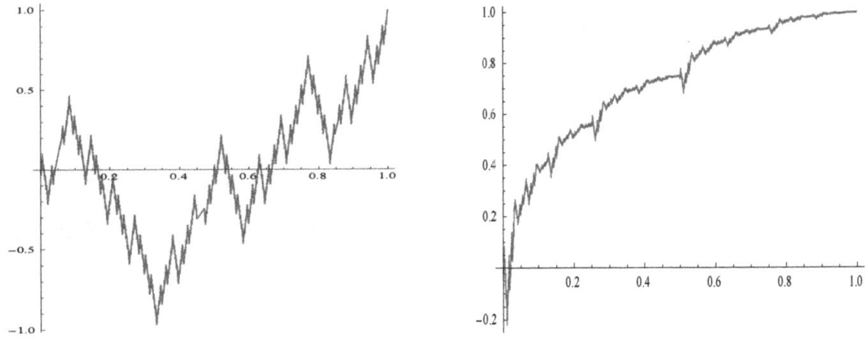

Fig. 2.4 The hybrid Kiesswetter-Casino attractor.

2.5 Quaternionic Fractal Interpolation

In this section, we extend fractal interpolation to a quaternionic setting. As quaternions from a non-commutative division algebra, the non-commutativity generates more intricate fractal patterns.

First, we give a short introduction to quaternions and present some of those properties that are relevant for the remainder of this section. The interested reader is referred to the literature on the subject, a short and subjective list of which is [7, 13, 14, 18, 28].

2.5.1 A Brief Introduction to Quaternions

Let $\{e_1, e_2, e_3\}$ be the canonical basis of the Euclidean vector space $\mathbb{R}^3$. We call $\{e_1, e_2, e_3\}$ imaginary units and require that the following multiplication rules hold:

$$e_1^2 = e_2^2 = e_3^2 = -1, \tag{2.36}$$

$$e_1e_2 = e_3 = -e_2e_1, \quad e_2e_3 = e_1 = -e_3e_2, \quad e_3e_1 = e_2 = -e_1e_3. \tag{2.37}$$

Remark 2.7 Note that (2.37) is equivalent to $e_1e_2e_3 = -1$.

A *real quaternion* q is then an expression of the form

$$q = a + \sum_{i=1}^{3} v_i e_i, \quad a, v_1, v_2, v_3 \in \mathbb{R}.$$

The addition and multiplication of two quaternions $q_1 = a + \sum_{i=1}^{3} v_i e_i$ and $q_2 = b + \sum_{i=1}^{3} w_i e_i$ is defined by

$$\begin{aligned} q_1 + q_2 &:= (a+b) + \sum_{i=1}^{3} (v_i + w_i) e_i \\ q_1 q_2 &:= (ab - v_1w_1 - v_2w_2 - v_3w_3) \\ &\quad + (aw_1 + bv_1 + v_2w_3 - v_3w_2)e_1 \\ &\quad + (aw_2 + bv_2 - v_1w_3 + v_3w_1)e_2 \\ &\quad + (aw_3 + bv_3 + v - 1w_2 - v_2w_1)e_3. \end{aligned}$$

Each quaternion $q = a + \sum\limits_{i=1}^{3} v_i e_i$ may be decomposed as $q = \mathrm{Sc}(q) + \mathrm{Vec}(q)$ where $\mathrm{Sc}(q) = a$ is the *scalar part* of q and $\mathrm{Vec}(q) = v = \sum\limits_{i=1}^{3} v_i e_i$ is the *vector part* of q. $v = \mathrm{Vec}(q)$ is also called a *quaternionic vector*.

The *conjugate* $\overline{q}$ of the real quaternion $q = a + v$ is the quaternion $\overline{q} = a - v$. Note that $q\overline{q} = \overline{q}q = a^2 + |v|^2 = a^2 + \sum\limits_{i=1}^{3} v_i^2$. Therefore, we can define a norm on $\mathbb{H}$ by setting

$$|q| := \sqrt{q\overline{q}}.$$

The inverse of a quaternion q is given by

$$q^{-1} = \frac{\overline{q}}{|q|^2}.$$

It is straight-forward to establish that the collection of all real quaternions

$$\mathbb{H} := \mathbb{H}_{\mathbb{R}} := \left\{ a + \sum_{i=1}^{3} v_i e_i \, : \, a, v_1, v_2, v_3 \in \mathbb{R} \right\},$$

is a four-dimensional associative normed division algebra over $\mathbb{R}$. Due to the multiplication rules (2.36) and (2.37), $\mathbb{H}$ is not commutative. We also note that $\mathbb{H}$, as already indicated above, is a four-dimensional vector space over $\mathbb{R}$ with basis $\{e_0, e_1, e_2, e_3\}$, where $e_0 := 1$.

Remark 2.8 We note that $\mathbb{H}$ and $\mathbb{R}^4$ are identical as *point sets* but differ in their algebraic structure. The vector space $\mathbb{R}^4$ is not an algebra whereas $\mathbb{H}$ is one, albeit non-commutative.

Remark 2.9 As we can define a norm and therefore a metric on $\mathbb{H}$, $\mathbb{H}$ becomes a topological space where open sets are defined via the metric. All other topological concepts such as limits, convergence, compactness, etc., then follow. $\mathbb{H}$ thus becomes a topological space and, moreover, also a complete metric space.

We also remark that if $v = \sum_{j=1}^{3} v_j e_j$ and $w = \sum_{j=1}^{3} w_j e_j$ are quaternionic vectors, then

$$vw = -\langle v, w \rangle + v \wedge w, \tag{2.38}$$

where $\langle v, w \rangle := \sum_{j=1}^{3} v_j w_j$ is the scalar product of v and w and

$$v \wedge w := (v_2 w_3 - v_3 w_2) e_1 + (v_3 w_1 - v_1 w_3) e_2 + (v_1 w_2 - v_2 w_1) e_3$$

is the vector (cross) product of v and w.

For our purposes, we need to introduce the analog of a Banach space in the quaternionic setting. To this end, we begin with the following definitions. (See, also [7].)

Definition 2.3 A real vector space V is called a left quaternionic vector space if it is a left $\mathbb{H}$-module, i.e., if there exists a mapping $\mathbb{H} \times \mathsf{V} \longrightarrow \mathsf{V}$, $(q, v) \mapsto qv$ which satisfies

1. $\forall v \in \mathsf{V}\, \forall q_1, q_2 \in \mathbb{H} : (q_1 + q_2) v = q_1 v + q_2 v$.
2. $\forall v_1, v_2 \in \mathsf{V}\, \forall q \in \mathbb{H} : q(v_1 + v_2) = q v_1 + q v_2$.
3. $\forall v \in \mathsf{V}\, \forall q_1, q_2 \in \mathbb{H} : q_1(q_2 v) = (q_1 q_2) v$.

Remark 2.10 In analogous fashion, one defines a *right* quaternionic vector space as a right $\mathbb{H}$-module where the mapping is now $\mathsf{V} \times \mathbb{H} \longrightarrow V$, $(v, q) \mapsto vq$.

A *two-sided quaternionic vector space* V is a left and right quaternionic vector space such that $\lambda v = v\lambda$, for all $\lambda \in \mathbb{R}$ and for all $v \in \mathsf{V}$. An example of a two-sided quaternionic vector space is given by $\mathbb{H}$ itself.

Remark 2.11 A quaternionic vector space becomes a real vector space when its scalar multiplication is restricted to $\mathbb{R}$.

One can start with any real vector space $\mathsf{V}_{\mathbb{R}}$ and construct a two-sided quaternionic vector space by setting

$$\mathsf{V}_{\mathbb{H}} := \left\{ \sum_{i=0}^{3} v_i \otimes e_i \, : \, v_i \in \mathsf{V}_{\mathbb{R}} \right\},$$

where $\otimes$ denotes the algebraic tensor product. (See, for instance, [30].)

On the other hand, given any two-sided quaternionic vector space and defining

$$\mathsf{V}_{\mathbb{R}} := \{v \in \mathsf{V} : \lambda v = v\lambda, \ \forall \lambda \in \mathbb{H}\},$$

then $\mathsf{V}_{\mathbb{R}}$ is a real vector space called the *real part of* V.

The proof of the next result can be found in, i.e., [30].

Proposition 2.4 *Let* V *be a two-sided quaternionic vector space and let* $\mathsf{V}_{\mathbb{R}}$ *denote its real part. Then* $V \cong \mathsf{V}_{\mathbb{R}} \otimes \mathbb{H}$.

Next, we introduce quaternionic normed spaces.

Definition 2.4 Let V be a left quaternionic vector space. A function $\|\cdot\| : \mathsf{V} \longrightarrow \mathbb{R}_0^+$ is called a norm on V if

1. $\|v\| = 0$ iff $v = 0$.
2. $\|qv\| = |q|\,\|v\|$, for all $v \in \mathsf{V}$ and $q \in \mathbb{H}$.
3. $\|v + w\| \leq \|v\| + \|w\|$, for all $v, w \in \mathsf{V}$.

A left quaternionic vector space endowed with a norm will be called a left quaternionic normed space.

A left quaternionic normed space E is called *complete* if it is a complete metric space with respect to the metric $d(x, y) = \|x - y\|$ induced by the norm $\|\cdot\|_{\mathsf{E}}$. In this case, we refer to E as *left quaternionic Banach space*.

Remark 2.12 A left (or right) quaternionic Banach space becomes a real Banach space if the left (right) scalar multiplication is restricted to $\mathbb{R}$. (Cf., [8, Section 2.3].)

Example 2.5 The space $\mathbb{H}^k$ consisting of k-tuples of quaternions is both a left and a right quaternionic vector space. We represent elements $\xi \in \mathbb{H}^k$ as column vectors and define the quaternionic conjugate * of ξ by

$$\begin{pmatrix} \xi_1 \\ \vdots \\ \xi_k \end{pmatrix}^* := \begin{pmatrix} \overline{\xi_1} & \cdots & \overline{\xi_k} \end{pmatrix},$$

where each $\xi_j \in \mathbb{H}$. When endowed with the norm

$$\|\xi\|_k := \sqrt{\xi^* \xi} = \sqrt{\sum_{l=1}^{k} |\xi_j|^2}, \tag{2.39}$$

$\mathbb{H}^k$ becomes a two-sided quaternionic Banach space as $\lambda v = v\lambda$, for all $\lambda \in \mathbb{R}$ and for all $v \in \mathbb{H}^k$.

Note that under the norm $\|\cdot\|_k$, $\mathbb{H}^k$ becomes a topological space and also a complete metric space.

The final concept we need to introduce is that of left linear mapping between left quaternionic vector spaces.

Definition 2.5 Let V_1 and V_2 be left quaternionic vector spaces. A mapping $f : \mathsf{V}_1 \longrightarrow \mathsf{V}_2$ is called left linear if

$$f(qv + w) = qf(v) + f(w), \quad \forall v, w \in \mathsf{V}, \forall q \in \mathbb{H}.$$

A left linear mapping is called bounded if

$$\|f\| := \sup_{x, y \in \mathsf{V}_1, x \neq y} \frac{\|f(x) - f(y)\|_{\mathsf{V}_2}}{\|x - y\|_{\mathsf{V}_1}} < \infty.$$

2.5.2 Quaternionic Fractal Interpolation

In this section, we introduce the novel concept of quaternionic fractal interpolation. For illustrative purposes, we do not choose the most general set up but restrict ourselves to the case where the left quaternionic Banach spaces are $\mathsf{E} := \mathbb{H}^k =: \mathsf{F}$, $k \in \mathbb{N}$.

For our purposes, we need the following function space. Let $\mathsf{X} \subset \mathbb{H}^k$ be compact (as defined via the norm $\|\cdot\|_k$) and let

$$\mathcal{B}(\mathsf{X}, \mathbb{H}^k) := \left\{ f : \mathsf{X} \to \mathbb{H}^k : f \text{ is bounded} \right\}.$$

A function $f : \mathsf{X} \to \mathbb{H}^k$ is called *bounded* if there exists a real number $M > 0$ such that $\|f\|_k \le M$. If we define for $x \in \mathsf{X}$ and $\lambda \in \mathbb{H}$

$$(f + g)(x) := f(x) + g(x) \quad \text{and} \quad (\lambda \cdot f)(x) := \lambda \cdot f(x)$$

then $\mathcal{B}(\mathsf{X}, \mathbb{H}^k)$ becomes a left quaternionic vector space. Setting for each $f \in \mathcal{B}(\mathsf{X}, \mathbb{H}^k)$

$$\|f\| := \sup_{x \in \mathsf{X}} \|f(x)\|_k,$$

then $\mathcal{B}(\mathsf{X}, \mathbb{H}^k)$ becomes a left Banach space. (See, for instance, [16, Chapter IV.E.1.] for the case $k = 1$. The extension to $k > 1$ is straight-forward.)

For the sake of simplicity and the purpose of understanding the underlying issues, we concentrate on the special case that

$$\mathsf{X} := \left\{ q \in \mathbb{H} : \max_{i=0,1,2,3} |q_i| = 1 \right\} \cong [-1, 1]^4$$

and that $k := 1$. The interested reader is encouraged to consider the general set-up.

We take the four-dimensional cube X and divide it into $n := 2^4$ congruent four-dimensional subcubes X_i each similar to X and such that $\{\mathsf{X}_i\}_{i=1}^n$ forms a partition of X in the sense of (2.1) and (2.2). We leave it to the diligent reader to derive closed expressions for the n injections l_i.

On the left Banach space $\mathcal{B}(\mathsf{X}, \mathbb{H})$, we consider the RB operator $T : \mathcal{B}(\mathsf{X}, \mathbb{H}) \to \mathcal{B}(\mathsf{X}, \mathbb{H})$ given by

$$Tf(l_i(x)) := q_i(x) + s_i(x) f(x), \quad x \in \mathsf{X}, \;\; i \in \mathbb{N}_n, \tag{2.40}$$

where $q_i, s_i : \mathsf{X} \to \mathbb{H}$ are bounded functions. Clearly, $Tf \in \mathcal{B}(\mathsf{X}, \mathbb{H})$ under these assumptions. Note that we left-multiply f by s_i.

Remark 2.13 In the case of real Banach spaces, the RB operator is affine, i.e., $T - T(0)$ is a linear operator. In the quaternionic setting this is no longer true: $T - T(0)$ is not a left linear operator. It is only if $s_i(x) \in \mathbb{R}$.

To simplify notation, we introduce the following abbreviation for the m-fold composition of functions from an IFS. Let $\mathcal{F} := \{f_1, \ldots, f_n\}$. We write

$$f_{i_m i_{m-1} \cdots i_1} := f_{i_m} \circ f_{i_{m-1}} \circ f_{i_1},$$

where each $i_j \in \mathbb{N}_n$.

For each $m \in \mathbb{N}$, the m-fold application of T, $T^m f := T(T^{m-1} f)$, can be written as

$$T^m f(l_{i_m i_{m-1} \cdots i_1}(x)) = \sum_{k=1}^{m} \prod_{j=1}^{k-1} s_{i_j}(x) q_k(x) + \prod_{k=1}^{m} s_{i_k}(x) f(x),$$

where the factors in the products $\prod$ are left-multiplied and where we set the empty product equal to 1. Notice the reverse order of the indices $i_m i_{m-1} \cdots i_1$ on the left- and right-hand side.

Continuing as in the previous section to ensure the existence of a fixed point ψ for the RB operator (2.40) or a bounded solution of the associated system of functional equations, we obtain in summary the next result.

Theorem 2.11 *For the above setting, the RB operator T defined in* (2.40) *has a unique fixed point $\psi \in \mathcal{B}(\mathsf{X}, \mathbb{H})$, i.e.,*

$$T\psi = \psi \iff \psi(l_i(x)) = q_i(x) + s_i(x)\psi(x), \quad x \in \mathsf{X}, \quad i \in \mathbb{N}_n, \tag{2.41}$$

provided that

$$\max_{i \in \mathbb{N}_n} \sup_{x \in \mathsf{X}_i} |s_i(x)| < 1.$$

The fixed point ψ is then called a *bounded quaternionic fractal function.*

Remark 2.14 A similar result exists of course for right Banach spaces and RB operators where the functions s_i are right-multiplied onto f. Note that the fixed point ψ depends on this right or left multiplication.

Example 2.6 This example shows the versatility of quaternionic fractal interpolation. We choose as $\mathsf{X} := \{q \in \mathbb{H} : \operatorname{Sc} q \in [0,1) \wedge \operatorname{Vec} q = 0\}$. Note that $\mathsf{X} \cong [0,1] \subset \mathbb{R}$. Define injections $l_i : \mathsf{X} \to \mathsf{X}$ as follows:

$$l_1(x) := \tfrac{1}{2}x \quad \text{and} \quad l_2(x) := \tfrac{1}{2}(x+1).$$

Clearly, $\{l_1(\mathsf{X}), l_2(\mathsf{X})\}$ is a partition of X satisfying (2.1) and (2.2).

Moreover, let $q_1 := e_0 + 2e_1 - e_3 + 3e_4$ and $q_2 := -e_0 - 2e_1 + 2e_3 + e_4$ be two quaternions. We set

$$q_1(x) := (1 - q_1)x \quad \text{and} \quad q_2(x) := q_2 x^2.$$

Then, $q_1, q_2 \in \mathcal{B}(\mathsf{X}, \mathbb{H})$. We define an RB operator T by

$$Tf(x) := \begin{cases} 2(1-q_1)x + s_1 f(2x), & x \in [0, \frac{1}{2}), \\ q_2(2x-1)^2 + s_2 f(2x-1), & x \in [\frac{1}{2}, 1), \end{cases}$$

where $s_1 := \frac{1}{10}e_0 + \frac{1}{2}e_1 - \frac{1}{5}e_2 - \frac{1}{10}e_3$ and $s_1 := -\frac{1}{5}e_0 + \frac{1}{5}e_1 - \frac{3}{5}e_2 + \frac{1}{10}e_3$. Note that $|s_1| = \frac{1}{10}\sqrt{31}$ and $|s_2| = \frac{1}{10}\sqrt{45}$. Thus, $\max\{s_1, s_2\} < 1$.

Therefore, the RB operator T is a contraction on $\mathcal{B}(\mathsf{X}, \mathbb{H})$ and has a unique fixed point $\psi \in \mathcal{B}(\mathsf{X}, \mathbb{H})$. The following Figure 2.5 shows the projection of the graph of ψ onto the (e_0, e_i)-plane for $i = 0, 1, 2, 3$.

As ψ can be written as $\psi = \sum\limits_{i=0}^{3} \psi_i e_i$, we display in Figure 2.6 the parametric plots (ψ_0, ψ_1, ψ_2) and (ψ_0, ψ_2, ψ_4).

2.6 Summary

In this chapter, we considered several aspects of fractal interpolation and identified their commonalities. Fractal interpolation is based either on solving a system of functional equations [31, 108] or on obtaining the fixed point of an operator [21, 22]. Both approaches yield the same results and are interchangeable.

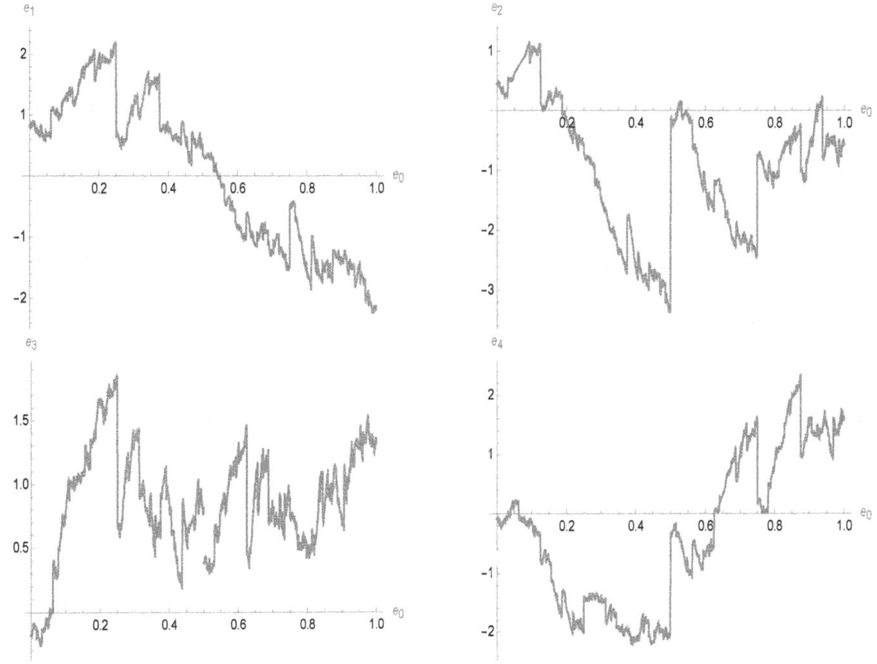

Fig. 2.5 The projections of ψ onto the (e_0, e_i)-planes.

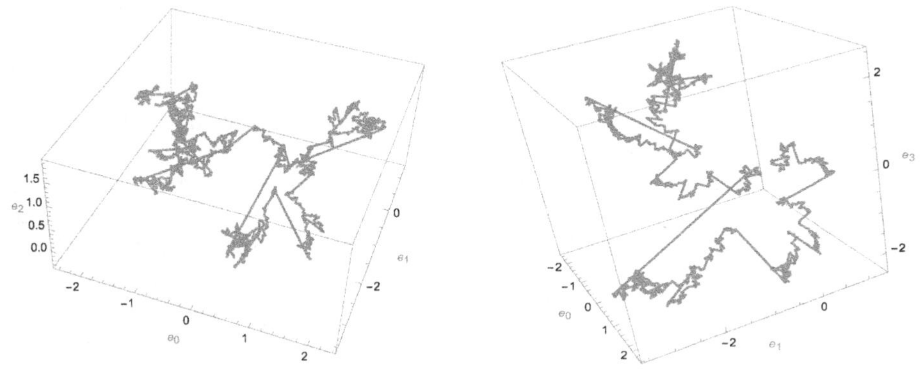

Fig. 2.6 Some parametric plots of the components of ψ.

In particular, we presented

- Global Fractal Interpolation: This was the original approach undertaken in [1] which is based on a geometric setting. The analytic setting based on an RB operator commenced in [6, 11, 12] and produced numerous generalizations and extensions. The main idea is to use a fixed bounded subset X of a normed space E to produce a partition of E over which copies of functions defined on the global set X reside.

Each of the partitioning sets is then iteratively partitioned into subsets thus producing the fractal nature of the limiting object.

- Local Fractal Interpolation: Here, we allow subsets of a bounded subset X of a normed space E to produce a partition of X. These subsets can be repeated in the construction of the partition and need no longer be defined by contractive injections. Here, functions defined on the local subsets reside over partitions of X. This type of local interpolation was successfully employed in fractal compression and image analysis. (Cf., for instance, [5].)
- Non-stationary Fractal Interpolation: In this approach, we no longer require that we keep the same quantities at each level of iteration but allow them to vary. This set-up is very similar to non-stationary subdivision and was investigated in this respect in [11, 16, 26]. This new methodology shows great potential for future investigations into the subject.
- Quaternionic Fractal Interpolation: In this novel approach, we kept the set-theoretic structure but allowed for a change in the algebraic structure of the underlying sets. By requiring that we are using a division algebra such as the quaternions or more generally hypercomplex numbers, we obtain even more flexibility in the construction of fractal interpolants. The investigation into this type of setting commenced in [27] and promises to be a fruitful field of investigation in years to come.

References

[1] Barnsley, M.F. 1986. Fractal functions and interpolation. Constr. Approx, 2: 303–329.

[2] Barnsley, M.F., B. Harding, C. Vince and P. Viswanathan. 2016. Approximation of rough functions. J. Approx, Th., 209: 23–43.

[3] Barnsley, M.F., M. Hegland and P.R. Massopust. 2014. Numerics and fractals. Bull. Inst. Math. Acad. Sin. (N.S.), 9(3): 389–430.

[4] Barnsley, M.F., M. Hegland and P.R. Massopust. Self-referential functions. https://arxiv.org/abs/1610.01369.

[5] Barnsley, M.F. and L.P. Hurd. 1993. Fractal Image Compression, AK Peters Ltd., Wellesly, MA.

[6] Bedford, T., M. Dekking and M. Keane. 1992. Fractal image coding techniques and contraction operators. Delft University of Technology Report, 92–93.

[7] Bourbaki, N. 1970. Éléments de mathématiques. Algèbre. Chapitres 1 á 3, Hermann, Paris.

[8] Colombo, F., J. Gantner and D.P. Kimsey. 2010. Spectral theory on the S-spectrum for quaternionic operators. Operator Theory: Applications and Advances, Vol. 270. Birkhäuser Verlag, Switzerland.

[9] Dira, N., D. Levin and P.R. Massopust. 2020. Attractors of trees of maps and of sequences of maps between spaces and applications to subdivision. J. Fixed Point Theory Appl., 22(14): 1–24.

[10] Dubins, L.E. and L.J. Savage. 1976. Inequalities for Stochastic Processes. Dover Publications: New York.

[11] Dubuc, S. 1986. Interpolation through an iterative scheme. J. Math. Anal. Appl., 114(1): 85–204.

[12] Dubuc, S. 1989. Interpolation fractale. *In*: Bélais, J. and S. Dubuc (eds.). Fractal Geometry and Analysis. Kluwer Academic Publishers, Dordrecht, The Netherlands.

[13] Gürlebeck, K., K. Habetha and W. Sprößig. 2000. Holomorphic Functions in the Plane and n-dimensional Space. Birkhäuser Verlag: Basel, Switzerland.

[14] Gürlebeck, K. and W. Sprößig. 1997. Quaternionic and Clifford Calculus for Physicists and Engineers. John Wiley & Sons: Chichester, England.

[15] Hutchinson, J.E. 1981. Fractals and self-similarity. Indiana Univ. Math. J., 30: 713–747.

[16] Jamison, J.E. 1970. Extension of some theorems of complex functional analysis to linear spaces over the quaternions and cayley numbers. Doctoral Dissertations, 2037, https://scholarsmine.mst.edu/doctoral_dissertations/2037.

[17] Kiesswetter, K. 1966. Ein einfaches Beispiel für eine Funktion welche überall stetig und nicht differenzierbar ist. Math. Phys. Semesterber, 13: 216–221.

[18] Kravchenko, V. 2003. Applied Quaternionic Analysis. Heldermann Verlag: Lemgo, Germany.

[19] Levin, D., N. Dyn and P. Viswanathan. 2019. Non-stationary versions of fixed-point theory, with applications to fractals and subdivision. J. Fixed Point Theory Appl., 21: 1–25.

[20] Massopust, P.R. 1997. Fractal functions and their applications. Chaos, Solitons and Fractals, 8(2): 171–190.

[21] Massopust, P.R. 2010. Interpolation and Approximation with Splines and Fractals. Oxford University Press: Oxford, USA.

[22] Massopust, P.R. 2016. Fractal Functions, Fractal Surfaces, and Wavelets. 2nd ed., Academic Press: San Diego, USA.

[23] Massopust, P.R. 2014. Local fractal functions and function spaces. Springer Proceedings in Mathematics & Statistics: Fractals, Wavelets and their Applications, 92: 245–270.

[24] Massopust, P.R. 2016. Local fractal functions in Besov and Triebel-Lizorkin spaces. J. Math. Anal. Appl., 436: 393–407.

[25] Massopust, P.R. 2016. Local fractal interpolation on unbounded domains. Proc. Edinburgh Math. Soc., 61: 151–167.

[26] Massopust, P.R. 2019. Non-stationary fractal interpolation. Mathematics, 7(8): 1–14.

[27] Massopust, P.R. 2019. Hypercomplex iterated function systems to appear in current trends in analysis, its applications and computation. *In*: Cereijeras, P., M. Reissig, I. Sabadini and J. Toft (eds.). Proceedings of the 12th ISAAC Congress, Birkhäuser, Aveiro, Portugal.

[28] Morais, J.P., S. Georgiev and W. Sprößig. 2014. Real Quaternionic Calculus, Birkhäuser Verlag: Basel, Switzerland.

[29] Navascués, M.A. 2005. Fractal polynomial interpolation. Z. Anal. Anwendungent, 24(2): 401–418.

[30] Ng, C. 2007. One quaternionic functional analysis. Math. Proc. Camb. Phil. Soc. 143: 391–406.

[31] Serpa, C. and J. Buesca. 2017. Constructive solutions for systems of iterative functional equations. Constructive Approx., 45(2): 273–299.

[32] Serpa, C. and J. Buesca. 2021. Compatibility conditions for systems of iterative functional equations with non-trivial contact sets. Results Math., 2: 1–19.

[33] Takagi, T. 1903. A simple example of the continuous function without derivative. Proc. Phys. Math. Soc. Japan, 1: 176–177.

Chapter 3

A Study on Fractal Operator Corresponding to Non-stationary Fractal Interpolation Functions

Saurabh Verma[1,*] and Sangita Jha[2]

3.1 Introduction

The notions of fractals naturally arise in the study of non-linear functions. The fractal theory provides alternative methods to describe, measure, and study natural phenomena like rivers, trees, lightning, heart sounds, mountains, etc. A useful method to construct fractals is by obtaining the fixed points of contractive operators for a particular type of Iterated Function Systems (IFSs) [13].

Fractal interpolation is an alternative to classical interpolation methods that form the basis of constructive approximation theory. The method is applied in several fields such as signal processing, computer-aided geometric design, image compression, and multiwavelets. Fractal Interpolation Function (FIF) is defined as a continuous function that interpolates a given set of data and whose graph is a fractal set. This notion was introduced by Barnsley in 1986 [3] and popularized by Navascués, Chand, and their colleagues [7, 8, 22, 23]. FIFs have more advantages than the classical interpolants for approximating and fitting the naturally occurring functions having self-similarity. Therefore, a large number of articles on this topic have been published so far. As an application point of view, recently, a new notion of dimension preserving approximation has been introduced in [30]. Further, this work is followed by [1] in the setting of bivariate fractal functions. For detailed discussions of FIFs and their advancement, we refer the reader to [1, 4, 8, 9, 19, 20, 31–33].

The fractal, defined as the attractor of a single IFS, is self-similar; its local shape is consistent under certain contraction maps. For the approximation of complicated curves with non-uniform self-similarity (for example, modeling Brownian motion and other stochastic processes), it is natural to think about the construction of fractals with different shapes at different levels. Barnsley et al. [5] considered an infinite family of IFSs to produce V-variable fractals. A V-variable fractal is a set that possesses at most V-distinct local patterns at each magnification level, where the set of patterns depends on the level. It is assumed that the sequence of IFSs is uniformly contractive and uniformly bounded. Recently, David and others [11, 16] introduced a more general class of sequences of operators and studied their convergence properties using the idea of a non-stationary subdivision algorithm. The authors in [11] have used the sequence of IFS systems in the field of non-stationary subdivision schemes. The uniform contractivity condition assumed in the construction of V-variable fractals is relaxed in this process. Further, the use of different contractive operators gives the flexibility to obtain the attractors (fractals) with different patterns at different scales. Recently, Massopust [21] studied the non-stationary FIF.

The IFS parameters have a crucial consequence on the properties and shapes of the corresponding FIFs [2, 14, 22]. Therefore, the intriguing question is identifying the IFS parameters and finding the fractal dimension of the non-stationary FIFs. The above problem is one of the most challenging questions in fractal

[1] Department of Applied Sciences, IIIT Allahabad, Prayagraj, India 211015.

[2] Department of Mathematics, NIT Rourkela, Rourkela, India 769008.

Email: sangitajha285@gmail.com

* Corresponding author: saurabhverma@iiita.ac.in

theory. The complexity of the above problem is further compounded by finding the exact upper bound of the box-counting and Hausdorff dimension. Some dimension aspects of fractal functions have been discussed in [15, 30].

The structure of the chapter is as follows. At first, we discuss the concept of the system of function systems and trajectories in Section 2. The construction of non-stationary univariate α-fractal function is described in Section 3. In Section 3.4, we study some properties of the associated fractal operator. Next, we attempt to find the bound of the box and Hausdorff dimension of the proposed interpolant in Section 3.5. In the end, we study the approximations of a continuously proposed interpolant fractal polynomials.

3.2 Backgrounds

We begin by recalling the basics of a sequence of function systems and backward trajectories of sequence of contraction maps, and refer the reader to [11, 23] for a more detailed exposition.

3.2.1 Sequences of Transformations and Trajectories

Let (X, d) be a complete metric space. For a map $f : X \to X$, let

$$\mathrm{Lip}(f) = \sup\left\{\frac{d(f(x), f(y))}{d(x, y)} : x, y \in X, x \neq y\right\}$$

denotes the Lipschitz constant of f. Then f is called Lipschitz if $\mathrm{Lip}(f) < \infty$, and a contraction if $\mathrm{Lip}(f) < 1$. We denote the collection of all non-empty compact subsets of X by $H(X)$ and define the Hausdorff distance between A and B in $H(X)$ as

$$h(A, B) = \max\{d(A, B), d(B, A)\}, \text{ where } d(A, B) = \sup_{x\in A}\inf_{y\in B} d(x, y).$$

The space $(H(X), h)$ is known as the space of fractals [13]. Let $\{w_i : X \to X; i = 1, 2, \ldots, N\}$ be a collection of continuous functions on X. Then the set $\{X; w_i : i = 1, 2, \ldots, N\}$ is called an IFS on X. If the maps $w_i, i = 1, 2, \ldots, N$ are contraction, then the above IFS is contractive. For a contractive IFS $\{X; w_i : i = 1, 2, \ldots, N\}$, the set valued Hutchinson map $W : H(X) \to H(X)$ is defined as

$$W(A) = \bigcup_{i=1}^{N} w_i(A).$$

It is easy to check that W is a contraction map on $H(X)$ with the Lipschitz constant $\mathrm{Lip}(W) = \max\{\mathrm{Lip}(w_i) : i = 1, 2, \ldots, N\}$. As a consequence of the Banach fixed point theorem, there exists a $G \in H(X)$ such that $G = W(G)$. This G is called the attractor or the deterministic fractal of the IFS (see [3] for details). The attractor G can be obtained as the limit of the iterative process $G_k = W(G_{k-1}),\ k \in \mathbb{N}$.

For the concept of trajectories, we consider a sequence of transformations $\{T_i\}_{i\in\mathbb{N}}$, $T_i : X \to X$.

Definition 3.1 (Invariant Set) A subset $\mathcal{P}$ of X is called an invariant set of the sequence $\{T_i\}_{i\in\mathbb{N}}$ if for all $i \in \mathbb{N}$ and $\forall\, x \in \mathcal{P}$, $T_i(x) \in \mathcal{P}$.

To obtain an invariant set from a sequence of transformations $\{T_i\}_{i\in\mathbb{N}}$, we shall review the following results from [16].

Lemma 3.1 *Consider a sequence of transformations* $\{T_i\}_{i\in\mathbb{N}}$ *on* (X, d). *Suppose there exists a* $y \in X$ *such that*

$$d(T_i(x), y) \leq cd(x, y) + M,$$

for all $x \in X$, $c \in [0, 1)$ *and* $M > 0$. *Then the ball* $B_r(y)$ *of radius* $r = \frac{M}{1-c}$ *centered at* y *is an invariant set for* $\{T_i\}_{i\in\mathbb{N}}$.

For a given complete metric space (X, d), we consider a sequence of set valued maps $W_k : H(X) \longrightarrow H(X)$ associated with Systems of Function Systems (SFS) as

$$W_k(A) = \bigcup_{i=1}^{n_k} w_{i,k}(A), A \in H(X), \tag{3.1}$$

where $W_k = \{w_{1,k}, w_{2,k}, \ldots, w_{n_k,k}\}$ is a family of contractions constituting an IFS on (X, d).

Definition 3.2 (Forward and Backward Trajectories) Let X be a metric space and $\{T_k\}_{k\in\mathbb{N}}$ be a sequence of Lipschitz maps on X. We define forward and backward procedures

$$\Phi_k := T_k \circ T_{k-1} \circ \ldots T_1 \text{ and } \Psi_k := T_1 \circ T_2 \circ \ldots T_k.$$

In [16], the authors studied the convergence of both types of trajectories and observed that the limits of forward trajectories do not lead to a new class of fractals. On the other hand, it is found that mild conditions are needed for the convergence of backward trajectories. Thus, we review the convergence results in the following. For proof and detailed investigations of the trajectories, the reader is asked to see [16].

Theorem 3.1 ([16], Corollary 4.4) *Let* $\{W_k\}_{k\in\mathbb{N}}$ *be a family of set-valued maps as described in* (3.1), *where the elements are collections* $W_k = \{w_{i,k} : i \in \mathbb{N}_{n_k}\}$ *of contractions on a complete metric space* (X, d). *Assume that*

1. *there exists a nonempty closed invariant set* $\mathcal{P} \subset X$ *for* $w_{i,k}, i \in \mathbb{N}_{n_k}, k \in \mathbb{N}$ *and*
2. $\sum_{k=1}^{\infty} \prod_{j=1}^{k} \text{Lip}(W_j) < \infty$.

Then the backward trajectories $\{\Psi_k(A)\}$ *converges for any initial* $A \subseteq \mathcal{P}$ *to a unique attractor* $G \subseteq \mathcal{P}$.

Remark 3.1 In view of Lemma 3.1, the authors of [16] mentioned that the first condition of the above theorem could be weaker. We remark that in [16, Section 4.1], the authors compared the above attractor with V-variable fractals [5] by showing that systems of function systems (SFSs) have weaker prerequisites than V-variable fractals.

3.3 Non-stationary Univariate α-fractal Functions

In this section, we construct the non-stationary univariate α-fractal function. For details on stationary α-fractal functions, see the following references [22, 23, 27, 29]. Let $I = [a, b]$ and $f : I \longrightarrow \mathbb{R}$ be a continuous function. Define a partition Δ by

$$\Delta : \{(x_1, \ldots, x_N) : a = x_0 < x_1 < \cdots < x_N = b\}.$$

For notational simplicity, we write $\mathbb{N}_N = \{1, 2, \dots, N\}$. Let $I_i := [x_{i-1}, x_i]$ for $i \in \mathbb{N}_N$, $P_i : I \to I_i$ be the affine map of the form $P_i(x) = a_i x + b_i$ and satisfy

$$P_i(x_1) = x_{i-1},\ P_i(x_N) = x_i,\ i \in \mathbb{N}_N. \tag{3.2}$$

Assume that $F_i : I \times \mathbb{R} \to \mathbb{R}$ is a function of the form

$$F_i(x, y) = \alpha_i(x) y + q_i(x),$$

where α_i are scaling functions and satisfy $\|\alpha\|_\infty = \max\{\|\alpha_i\|_\infty : i \in \mathbb{N}_N\} < 1$. The maps $q_i : I \to \mathbb{R}$ are continuous functions satisfying the conditions:

$$F_i(x_1, y_1) = y_{i-1}, \quad F_i(x_N, y_N) = y_i, \quad i \in \mathbb{N}_N. \tag{3.3}$$

The system $\mathcal{I} = \{I \times K; w_i(x, y) : i \in \mathbb{N}_N\}$, where $w_i(x, y) := (P_i(x), F_i(x, y))$ is a IFS. Using Theorem 1 of [3], one can show that the IFS $\mathcal{I}$ has a unique attractor G and it is the graph of a continuous function that interpolates the data $\{(x_i, y_i) : i = 0, 1, 2, \dots, N\}$. The corresponding interpolating function is called a fractal interpolation function. With these preliminaries, we are now ready to provide the construction of the non-stationary α-fractal FIF. Set $K = I \times \mathbb{R}$ and define $F_{i,k} : K \to \mathbb{R}$ by

$$F_{i,k}(x, y) = \alpha_{i,k}(x) y + f\big(P_i(x)\big) - \alpha_{i,k}(x)) u_k(x),$$

where $\alpha_{i,k} : I \to \mathbb{R}$ are continuous functions such that

$$\|S\|_\infty := \sup_{k \in \mathbb{N}}\{\|\alpha_{i,k}\|_\infty : i \in \mathbb{N}_N\} < 1.$$

Let $u_k \in C(I, \mathbb{R})$ be a function satisfying $u_k \neq f$, $\|u\|_\infty := \sup_{k \in \mathbb{N}} \|u_k\|_\infty < \infty$,

$$u_k(x_i) = f(x_i) \quad \text{for } i = 0, N.$$

Let us mention two examples for such function $u_k \in C(I, \mathbb{R})$.

1. $u_k(x) = f(x) t_k(x)$, where $t_k \in C(I, \mathbb{R})$ is a fixed non-constant function such that $t_k(x_i) = 1$ for $i = 0, N$.
2. $u_k(x) = (f \circ t_k)(x)$, where $t_k \in C(I, I)$ is a fixed map $t_k \neq Id$, the identity map and $t_k(x_i) = (x_i)$ for $i = 0, N$.

For each $i \in \mathbb{N}_N$, we define $W_{i,k} : K \to I_i \times \mathbb{R}$ by

$$W_{i,k}(x, y) = \big(P_i(x), F_{i,k}(x, y)\big),$$

which forms a sequence of IFSs $\mathcal{I}_k := \big\{K; W_{i,k} : i \in \mathbb{N}_N\big\}$.

Definition 3.3 Two sequences $\{x_k\}_{k \in \mathbb{N}}$ and $\{y_k\}_{k \in \mathbb{N}}$ in a metric space (X, d) are said to be asymptotically similar if $d(x_k, y_k) \to 0$ as $k \to \infty$.

Example 3.1 Let $X = \mathbb{R}$ and $x_k = \frac{1}{k}$, $y_k = \frac{1}{2k}$. Then $\{x_k\}_{k \in \mathbb{N}}$ and $\{y_k\}_{k \in \mathbb{N}}$ are asymptotically similar.

Remark 3.2 If $\{x_k\}_{k \in \mathbb{N}}$ and $\{y_k\}_{k \in \mathbb{N}}$ are asymptotically similar then

$$\lim_{k \to \infty} x_k = x \iff \lim_{k \to \infty} y_k = x.$$

Proposition 3.1 *Let $\{T_k\}_{k\in\mathbb{N}}$ be a sequence of Lipschitz maps on a complete metric space X such that T_k has Lipschitz constant c_k. If $\lim_{k\to\infty}\prod_{i=1}^{k} c_i = 0$, then $\{\Phi_k(x)\}$, $\{\Phi_k(y)\}$ are asymptotically similar for all $x, y \in X$, and so are $\{\Psi_k(x)\}$, $\{\Psi_k(y)\}$ for all $x, y \in X$.*

The following theorem is borrowed from [24]. We include the proof for completeness and record.

Proposition 3.2 *Let $\{T_k\}_{k\in\mathbb{N}}$ be a sequence of Lipschitz maps on a complete metric space X. If there exists $x_* \in X$ such that the sequence $\{d(x_*, T_k(x_*))\}$ is bounded, and $\sum_{k=1}^{\infty}\prod_{i=1}^{k} c_i < \infty$ then the sequence $\{\Psi_k(x)\}$ converges for all $x \in X$ to a unique limit $\overline{x}$.*

Proof For $m, k \in \mathbb{N}$, $m > k$, and $x_* \in X$ satisfying the assumption of the proposition, we get

$$\begin{aligned} d(\Psi_m(x_*), \Psi_k(x_*)) \leq\, & d(\Psi_m(x_*), \Psi_{m-1}(x_*)) + d(\Psi_{m-1}(x_*), \Psi_{m-2}(x_*)) + \\ & \cdots + d(\Psi_{k+2}(x_*), \Psi_{k+1}(x_*)) + d(\Psi_{k+1}(x_*), \Psi_k(x_*)) \\ \leq & \Big(\prod_{i=1}^{m-1} c_i\Big) d(T_m(x_*), x_*) + \cdots + \Big(\prod_{i=1}^{k} c_i\Big) d(T_{k+1}(x_*), x_*). \end{aligned}$$

Now, the condition of boundedness of the sequence $\{d(x_*, T_i(x_*))\}$ infer that for some $M > 0$,

$$d(\Psi_m(x_*), \Psi_k(x_*)) \leq M(S_{m-1} - S_{k-1}),$$

where $S_j = \sum_{k=1}^{j}\prod_{i=1}^{k} c_i$. This in turn yields that $\{\Psi_k(x)\}$ is Cauchy and therefore convergent. Using Proposition 3.1, all the trajectories $\{\Psi_k(y)\}$ are convergent to the same limit for all $y \in X$. This completes the proof. □

Remark 3.3 The above result can be compared with Proposition 3.11 of [16], wherein one additional assumption of the compact invariant domain is considered. Further, in Remark 3.12 of [16], it is noted that the result of [16, Proposition 3.11] is true with the weaker assumption that C is an invariant domain of $\{T_k\}_{k\geq n_0}$ for some $n_0 \in \mathbb{N}$.

Remark 3.4 The conditions for convergence of the forward and backward trajectories are more general versions of [16] in [11]. In view of this, one can generalize our result to the setting of [11].

Consider the following notation:

$$\alpha_k = (\alpha_{1,k}, \ldots, \alpha_{N,k}),\ S := \{\alpha_k\}_{k\in\mathbb{N}} \text{ and } u := \{u_k\}_{k\in\mathbb{N}}.$$

Now, let $C_f(I) := \{g \in C(I) : g(x_i) = f(x_i)\ \forall\, i = 0, N\}$, it is obvious that $C_f(I)$ is a complete metric space. For $k \in \mathbb{N}$, we define a sequence of RB operators $T^{\alpha_k} : C_f(I) \to C_f(I)$ by

$$(T^{\alpha_k} g)(x) = F_{i,k}(P_i^{-1}(x), g(P_i^{-1}(x)))\quad \forall x \in I_i,\ i \in \mathbb{N}_N.$$

For completeness, we shall hint at the well-definedness of T^{α_k}.

Proposition 3.3 *The above $T^{\alpha_k} : C_f(I) \to C_f(I)$ is well-defined for each $k \in \mathbb{N}$.*

Proof It is obvious. □

We further note that T^{α_k} is a contraction map, that is,

$$\|T^{\alpha_k}g - T^{\alpha_k}h\|_\infty \le \|\alpha_k\|_\infty \|g-h\|_\infty,$$

where $\|\alpha_k\|_\infty = \max\{\|\alpha_{i,k}\|_\infty : i \in \mathbb{N}_N\}$. Now, we are ready to prove the non-stationary version of [[28], Theorem 3.1].

Theorem 3.2 *Let us consider the sequence of operators $\{T^{\alpha_k}\}$ on $C_f(I)$ defined above with the conditions described. Then for every $g \in C_f(I)$ the sequence $\{T^{\alpha_1} \circ T^{\alpha_2} \circ \cdots \circ T^{\alpha_k} g\}$ converges to a map of $C_f(I)$.*

Proof Let us consider $g \in C_f(I)$. We first check that $\{\|T^{\alpha_k}g - g\|_\infty\}$ is bounded. Applying the definition of T^{α_k},

$$\begin{aligned}\|T^{\alpha_k}g - g\|_\infty &\le \|f\|_\infty + \|g\|_\infty + \|\alpha_k\|_\infty(\|g\|_\infty + \|u_k\|_\infty)\\ &= (1+\|S\|_\infty)\|g\|_\infty + \|f\|_\infty + \|S\|_\infty\|u\|_\infty,\end{aligned}$$

the bound does not depend on k. Applying Proposition 3.2, there exists $f_u^S \in C_f(I)$ such that $f_u^S = \lim_{k\to\infty} T^{\alpha_1} \circ T^{\alpha_2} \circ \cdots \circ T^{\alpha_k} g$ for any $g \in C_f(I)$. □

Remark 3.5 The above theorem should be compared with [21, Theorem 4]. More precisely, [21, Theorem 4] uses the assumption of a closed invariant set (see, [21, Proposition 3]). However, our proof does not require that assumption. Hence, our result can be treated as a generalization of [21, Theorem 4].

Definition 3.4 The function f_u^S is the non-stationary α-fractal function with respect to f, S, u, and the partition Δ.

Remark 3.6 Since each T_k is a contraction, there exists a unique function f_k^α such that $T^{\alpha_k}(f_k^\alpha) = f_k^\alpha$. Being the fixed point of the RB-operator T^{α_k} [28], f_k^α satisfies the functional equation:

$$f_k^\alpha(x) = F_{i,k}\Big(Q_i(x), f_k^\alpha(Q_i(x))\Big) \quad \forall\ x \in I_i,$$

where $Q_i(x) := P_i^{-1}(x)$. That is, for all $i \in \mathbb{N}_N$ and $x \in I_i$, we have

$$f_k^\alpha(x) = f(x) + \alpha_k(x) f_k^\alpha(Q_i(x)) - \alpha_k(x) u_k(Q_i(x)). \tag{3.4}$$

3.4 Associated Non-stationary Fractal Operator on $C(I)$

Let us recall the notations from the previous section. Let $\|S\|_\infty = \sup_{k\in\mathbb{N}} \|\alpha_k\|_\infty < 1$ and $\|u\|_\infty = \sup_{k\in\mathbb{N}} \|u_k\|_\infty$. Further, assume that $u_k = L_k f$, where $L_k : C(I) \to C(I)$ is a bounded linear operators such that $L_k f(x_i) = f(x_i)\ for\ i \in \{0, N\},\ k \in \mathbb{N}$ and $\|L\| := \sup_{k\in\mathbb{N}} \|L_k\| < \infty$.

Definition 3.5 Let Δ, α_k and L_k be fixed. For each fixed $f \in C(I)$ to its perturbed function f_u^S, we obtain the non-stationary α-fractal operator $\mathcal{F}_u^S$ defined as

$$\mathcal{F}_u^S : C(I) \to C(I),\ \mathcal{F}_u^S(f) = f_u^S.$$

We now discuss some inequalities for the error in the process of approximating f with its non-stationary α-fractal analogue.

Theorem 3.3 *We have the following observations:*

(1) If S is the null sequence then $f_u^S = f$.
(2) The perturbation error is given by:

$$\|f_u^S - f\|_\infty \leq \sum_{k=1}^{\infty} \|S\|_\infty^k \|f - u_k\|_\infty \leq \frac{\|S\|_\infty}{1 - \|S\|_\infty} K_{f,u},$$

where $K_{f,u} = \sup_{k \in \mathbb{N}} \{\|f - u_k\|_\infty\}$.

(3) If the sequence $\{\alpha^m = \{\alpha_{i,k}^m : i \in \mathbb{N}_N\}_{m \in \mathbb{N}}$ converges to the null sequence in sup-norm then $f_u^{S^m}$ converges to f.

(4) If the sequence $\{u^m = \{u_k^m\}\}_{m \in \mathbb{N}}$ is such that $K_{f,u^m} \to 0$, then $f_u^{S^m}$ converges to f.

Proof (1) Recall that

$$T^{\alpha_k} g(x) = f(x) + \alpha_{i,k}(x) g(Q_i(x)) - \alpha_{i,k}(Q_i(x)) u_k(Q_i(x)),$$

for all $x \in I_i$, $i \in \mathbb{N}_N$. Let S be a null sequence. That is, $\alpha_{i,k} = 0$ for all $k \in \mathbb{N}$. Then we get $T^{\alpha_k} g(x) = f(x)$ for all $x \in I$. This on induction provides

$$T^{\alpha_1} \circ T^{\alpha_2} \circ \cdots \circ T^{\alpha_k} g(x) = f(x), \quad \forall\, x \in I.$$

On taking the limit $k \to \infty$, we have

$$f_u^S(x) = \lim_{k \to \infty} T^{\alpha_1} \circ T^{\alpha_2} \circ \cdots \circ T^{\alpha_k} g(x) = f(x), \quad \forall\, x \in I,$$

proving the assertion.

(2) By definition of RB operators T^{α_k} we have

$$T^{\alpha_1} \circ T^{\alpha_2} \circ \cdots \circ T^{\alpha_k} f(x) - f(x) = \alpha_1(x) \Big(T^{\alpha_2} \circ T^{\alpha_3} \circ \cdots \circ T^{\alpha_k} f - u_1\Big)(Q_i(x)),$$

for all $x \in I_i$. Inductively, we get

$$T^{\alpha_1} \circ T^{\alpha_2} \circ \cdots \circ T^{\alpha_k} f(x) - f(x) = \sum_{l=1}^{k} \alpha_1(x) \ldots \alpha_l(Q_i^l(x))(f - u_l)(Q_i^l(x)),$$

for all $x \in I_i$, where Q_i^l is a suitable finite composition of mappings Q_i. Now,

$$\|T^{\alpha_1} \circ T^{\alpha_2} \circ \cdots \circ T^{\alpha_k} f - f\|_\infty \leq \sum_{l=1}^{k} \|S\|_\infty^l \|f - u_l\|_\infty.$$

As $k \to \infty$, we have

$$\|f_u^S - f\|_\infty \leq \sum_{l=1}^{\infty} \|S\|_\infty^l \|f - u_l f\|_\infty \leq \frac{\|S\|_\infty}{1 - \|S\|_\infty} \sup_{k \in \mathbb{N}} \|f - u_k\|_\infty.$$

This completes the proof of item (2).

(3) For a sequence of scaling functions, from the above expression we obtain

$$\|f_u^{S^m} - f\|_\infty \le \frac{\|\alpha^m\|_\infty}{1 - \|\alpha^m\|_\infty} K_{f,u^m}.$$

Since $\|\alpha^m\|_\infty = \sup_{k\in\mathbb{N}}\{\|\alpha_k^m - 0\|_\infty\} \to 0$, we obtain $f_u^{S^m}$ converges to f.

(4) The proof is obtained using the result from item (2) and lines similar to item (3). □

Theorem 3.4 *The fractal operator $\mathcal{F}_u^S : C(I) \to C(I)$ is a bounded linear operator.*

Proof Let $f, g \in C(I)$ and $\beta, \gamma \in \mathbb{R}$. Using the proof of item (2) in Theorem 3.3, we can write

$$(\beta f)_u^S(x) = \beta f(x) + \lim_{k\to\infty} \sum_{l=1}^{k} \alpha_1 \dots \alpha_l(Q_i^l(x))\big(\beta f - L_l(\beta f)\big)(Q_i^l(x)),$$

and

$$(\gamma g)_u^S(x) = \gamma g(x) + \lim_{k\to\infty} \sum_{l=1}^{k} \alpha_1 \dots \alpha_l(Q_i^l(x))\big(\gamma g - L_l(\gamma g)\big)(Q_i^l(x)).$$

Using linearity of L_k, the above equations give

$$(\beta f)_u^S(x) + (\gamma g)_u^S(x) = (\beta f + \gamma g)(x) + \lim_{k\to\infty} \sum_{l=1}^{k} \alpha_1 \dots \alpha_l(Q_i^l(x))\big(\beta f + \gamma g - L_l(\beta f + \gamma g)\big)(Q_i^l(x)).$$

Now, since

$$(\beta f + \gamma g)_u^S(x) = (\beta f + \gamma g)(x) + \lim_{k\to\infty} \sum_{l=1}^{k} \alpha_1 \dots \alpha_l(Q_i^l(x))\big(\beta f + \gamma g - L_l(\beta f + \gamma g)\big)(Q_i^l(x)),$$

we deduce that

$$(\beta f + \gamma g)_u^S = (\beta f)_u^S + (\gamma g)_u^S,$$

that is, $\mathcal{F}_u^S$ is a linear operator. Again by item (2) of Theorem 3.3, we get

$$\|\mathcal{F}_u^S(f)\|_\infty - \|f\|_\infty \le \frac{\|S\|_\infty}{1 - \|S\|_\infty} \sup_{k\in\mathbb{N}} \|f - L_k f\|_\infty \le \frac{\|S\|_\infty}{1 - \|S\|_\infty} K_L \|f\|_\infty, \tag{3.5}$$

where $K_L = \sup_{k\in\mathbb{N}} \|Id - L_k\|$. Therefore,

$$\|\mathcal{F}_u^S(f)\|_\infty \le \left(1 + \frac{K_L\|S\|_\infty}{1 - \|S\|_\infty}\right)\|f\|_\infty.$$

Now we recall a few fundamental lemmas and definitions from [18].

Lemma 3.2 *Let X be a Banach space and $T : X \to X$ be a bounded linear operator. If $\|T\| < 1$, then $(Id - T)^{-1}$ exists and it is bounded.*

Lemma 3.3 *Let X be normed linear space. If $T_1 : X \to X$ is a bounded linear operator, and $T_2 : X \to X$ is a compact operator, then T_1T_2 and T_2T_1 are compact operators.*

Let X, Y be normed spaces and $T : X \longrightarrow Y$ be a bounded linear operator. The adjoint or dual T^* of T is the unique map $T^* : Y^* \longrightarrow X^*$ such that

$$\langle x, T^* g\rangle = \langle Tx, g\rangle, \ \ \forall\, x \in X, \ \ g \in Y^*.$$

Definition 3.6 An operator T is called Fredholm if: Range(T) is closed and $\ker(T)$, $\ker(T^*)$ are finite-dimensional.

Definition 3.7 We define the index of a Fredholm operator as

$$\text{index}(T) = \dim(\ker(T)) - \dim(\ker(T^*)).$$

Now we discuss some properties of the non-stationary α-fractal function and the corresponding fractal operator.

Theorem 3.5 *Let* $\|S\|_\infty = \sup_{k\in\mathbb{N}} \|\alpha_k\|_\infty < 1$, *and let* Id *be the identity operator on* $C(I)$.

1. *Let* f_u^S *be the non-stationary* α*-fractal function associated with a given continuous function* f, *partition* Δ, *and the linear operators* L_k. *Then we have the following inequality*
$$\|f_u^S - L_k f\|_\infty \leq \frac{1}{1 - \|S\|_\infty} \sup_{k\in\mathbb{N}} \|f - L_k f\|_\infty.$$
2. $\mathcal{F}_u^S$ *is bounded below and has closed range if the scaling functions satisfy* $\|S\|_\infty < \frac{1}{K_L}$.
3. $\mathcal{F}_u^S$ *is a topological automorphism on* $C(I)$ *if the scaling sequence satisfies* $\|S\|_\infty < \frac{1}{1+K_L}$.
4. *For* $\|S\|_\infty \neq 0$, *the fixed points of* L_k *coincides with the fixed points of* $\mathcal{F}_u^S$.
5. *If* 1 *belongs to the point spectrum of* L_k, *then* $1 \leq \|\mathcal{F}_u^S\|$.
6. *For* $\|S\|_\infty < \frac{1}{K_L}$, *the fractal operator* $\mathcal{F}_u^S$ *is not a compact operator.*
7. *If* $\|S\|_\infty < \frac{1}{1+K_L}$, *then* $\mathcal{F}_u^S$ *is Fredholm and its index is* 0.

Proof 1. We have $\|f_u^S - L_k f\|_\infty = \|f_u^S - f + f - L_k f\|_\infty$. From item (2) of Theorem 3.3, we obtain

$$\begin{aligned}\|f_u^S - L_k f\|_\infty &\leq \|f_u^S - f\| + \|f - L_k f\|_\infty \\ &\leq \frac{\|S\|_\infty}{1 - \|S\|_\infty} \sup_{k\in\mathbb{N}} \|f - L_k f\|_\infty + \sup_{k\in\mathbb{N}} \|f - L_k f\|_\infty.\end{aligned}$$

Thus we obtain,

$$\|f_u^S - L_k f\|_\infty \leq \frac{1}{1 - \|\alpha\|_\infty} \sup_{k\in\mathbb{N}} \|f - L_k f\|_\infty.$$

2. From equation (3.5), we obtain

$$\|f\|_\infty - \|f_u^S\|_\infty \leq \|f - f_u^S\|_\infty \leq \left(1 + \frac{K_L \|S\|_\infty}{1 - \|S\|_\infty}\right) \|f\|_\infty,$$

where K_L is prescribed in Theorem 3.4. Since $\|\alpha\|_\infty < \frac{1}{K_L}$, we get

$$\|f\|_\infty \leq \frac{1 - (1 + K_L)\|\alpha\|_\infty}{1 - \|\alpha\|_\infty K_L} \|f_u^S\|_\infty.$$

Thus $\mathcal{F}_u^S$ is bounded below. Now to show $\mathcal{F}_u^S$ has a closed range, let $\{f_{n,u}^S\}$ be a sequence in $\mathcal{F}_u^S(C(I))$ such that $f_{n,u}^S \to g$, i.e., $f_{n,u}^S$ is Cauchy in $C(I)$. Then from

$$\|f_n - f_m\|_\infty \le \left(1 + \frac{K_L\|S\|_\infty}{1-\|S\|_\infty}\right)\|f_{n,u}^S - f_{m,u}^S\|_\infty,$$

f_n is a Cauchy sequence in $C(I)$. Thus there exists $f \in C(I)$ such that $f_n \to f$. Using the continuity of $\mathcal{F}_u^S$, we obtain $g = \mathcal{F}_u^S = f_{n,u}^S$.

3. From the bound of Theorem 3.4, we can easily obtain

$$\|Id - \mathcal{F}_u^S\| \le \frac{\|S\|_\infty}{1-\|S\|_\infty}K_L,$$

where Id is the identity operator on $C(I)$. Using the assumption on the scaling sequence, we have $\|Id-\mathcal{F}_u^S\| < 1$. Consequently, $\sum_{j=0}^{\infty}(Id-\mathcal{F}_u^S)^j$ is convergent in the operator norm and $\mathcal{F}_u^S = Id-(Id-\mathcal{F}_u^S)$ is invertible.

4. Let $\|\alpha\|_\infty \neq 0$, and f be a fixed point of L_k.

$$\|f_u^S - f\|_\infty \le \|S\|_\infty \; \|f_u^S - f\|_\infty.$$

Since $\|S\|_\infty < 1$, this implies that $f_u^S = f$.

5. Consider $h \in C(I)$ with $\|g\| = 1$ and $L_k h = h$. Then, item (4) of Theorem 3.5 gives $\mathcal{F}_u^S(g) = g$. Consequently, $\|\mathcal{F}_u^S(g)\|_\infty = \|g\|_\infty$. The definition of the operator norm now yields $1 \le \|\mathcal{F}_u^S\|_\infty$.
6. For $\|\alpha\|_\infty < \frac{1}{K_L}$, we know that $\mathcal{F}_u^S : C(I) \to C(I)$ is one-one. Note that the range space of $\mathcal{F}_u^S$ is infinite dimensional. We consider the inverse map $(\mathcal{F}_u^S)^{-1} : \mathcal{F}_u^S(C(I)) \to C(I)$. For this α, $\mathcal{F}_u^S$ is bounded below, and hence it follows that $(\mathcal{F}_u^S)^{-1}$ is a bounded linear operator. Assume that $\mathcal{F}_u^S$ is a compact operator. Then by Lemma 3.3, we deduce that the operator $T = \mathcal{F}_u^S(\mathcal{F}_u^S)^{-1} : \mathcal{F}_u^S(C(I)) \to C(I)$ is a compact operator, which is a contradiction to the infinite dimensionality of the space $\mathcal{F}_u^S(C(I))$. Therefore, $\mathcal{F}_u^S$ is not a compact operator.
7. Under the hypothesis, range space of $\mathcal{F}_u^S$ is closed. Furthermore, $\mathcal{F}_u^S$ is invertible. It is known that if $T : X \to Y$ is invertible, then T^* is also invertible [6]. Therefore $(\mathcal{F}_u^S)^*$ is invertible. As a consequence, $\mathcal{F}_u^S$ is Fredholm. The index of a Fredholm operator is defined as

$$\text{index}(\mathcal{F}_u^S) = \dim\left(\ker(\mathcal{F}_u^S)\right) - \dim\left(\ker\,(\mathcal{F}_u^S)^*\right).$$

Hence, the index is zero.

In the next result we show the existence of a non-trivial closed invariant subspace for the fractal operator. The proof of this next theorem is given by modifying and adapting some standard techniques present in the literature on invariant subspace problem; see, for instance, [6].

Theorem 3.6 *There exists a non-trivial closed invariant subspace for the fractal operator* $\mathcal{F}_L^S : C(I) \to C(I)$.

Proof We consider a non-zero continuous function $f : I \to \mathbb{R}$ such that $f(x_i) = 0$ for every $x_i \in \Delta$. Denote by $(\mathcal{F}_L^S)^r$ the r-fold composition of $\mathcal{F}_L^S$ with itself, and $(\mathcal{F}_L^S)^0 := f$. Now, we construct the $\mathcal{F}_L^S$-cyclic subspace generated by f, that is,

$$Y_f = \text{span}\{f, \mathcal{F}_L^S(f), (\mathcal{F}_L^S)^2(f), \dots\}.$$

It is obvious that $Y_f \neq \{0\}$ and $\mathcal{F}_L^S(Y_f) \subseteq Y_f$. Let $g \in Y_f$. Using the definition of Y_f, there exist constants $\beta_i \in \mathbb{R}$ and $r_i \in \mathbb{N} \cup \{0\}$ such that

$$g = \beta_1(\mathcal{F}_L^S)^{r_1}(f) + \beta_1(\mathcal{F}_L^S)^{r_2}(f) + \cdots + \beta_m(\mathcal{F}_L^S)^{r_m}(f).$$

With the help of the interpolatory property of the fractal operator, we obtain $f(x_i) = (\mathcal{F}_L^S(f))(x_i)$ for every $(x_i) \in \Delta$. This implies that $g(x_i) = 0 \ \forall \ x_i \in \Delta$. Assume that $Y = \overline{Y_f}$. We immediately establish that $\mathcal{F}_L^S(Y) \subseteq Y$, hence that Y is a closed invariant subspace of $\mathcal{F}_L^S$.

To prove Y is a nontrivial closed invariant subspace of $\mathcal{F}_L^S$, we proceed as follows. Let $h \in Y$. Then there exists a sequence $(h_n)_{n\in\mathbb{N}} \subset Y_f$ such that $h_n \to h$ uniformly. Since uniform convergence implies pointwise convergence, we deduce that $h(x_i) = 0 \ \forall \ x_i \in \Delta$. From which we conclude that a continuous function that is nonzero at some points in Δ does not belong to Y. In particular, $Y \neq C(I)$, completing the proof. □

3.5 Fractal Dimension of Non-stationary α-fractal Functions

A function $f : I \to \mathbb{R}$ is said to be Hölder continuous with exponent σ if

$$|f(x) - f(y)| \leq k_f |x - y|^{\sigma} \quad \forall \, x, y \in I,$$

and for some $k_f > 0$.
For Hölder continuous functions f with exponent σ, let us define σth Hölder seminorm as

$$[f]_{\sigma} = \sup_{x \neq y} \frac{|f(x) - f(y)|}{|x - y|^{\sigma}}.$$

Consider the Hölder space

$$\mathcal{H}^{\sigma}(I) := \{g : I \to \mathbb{R} : \text{ g is Hölder continuous with exponent } \sigma\}.$$

The space $\mathcal{H}^{\sigma}(I)$ is a Banach space when endowed with the norm $\|g\|_{\sigma} := \|g\|_{\infty} + [g]_{\sigma}$.

We refer the reader to [17] for definitions and some properties of box dimension.

Remark 3.7 If $f \in \mathcal{H}^{\sigma_0}(I)$, then $f \in \mathcal{H}^{\sigma}(I)$ for each $0 < \sigma < \sigma_0$.

Let $\mathcal{H}_f^{\sigma}(I) = \{g \in \mathcal{H}^{\sigma}(I) : g(x_i) = f(x_i) \, \forall \, i = 0, N\}$. It is easy to check that $\mathcal{H}_f^{\sigma}(I)$ is closed subspace of $\mathcal{H}^{\sigma}(I)$. Now, we are ready to prove the next result.

Theorem 3.7 *Let f and $\alpha_{i,k}$ be Hölder continuous with exponent σ_1 and σ_2 respectively for every $k \in \mathbb{N}$. Let u_k be Hölder continuous with exponent σ_3 satisfying $u_k(x_i) = f(x_i)$ for $i = 0, N, \ k \in \mathbb{N}$. If $\max\left\{\|S\|_{\sigma}, \frac{\|S\|_{\infty}}{(\min\{|a_i|\})^{\sigma}}\right\} < 1$ then for any $g \in \mathcal{H}_f^{\sigma}(I)$, the sequence$\{\Psi_k(g)\}$ converges in norm $\|\cdot\|_{\sigma}$ to $f_u^S \in \mathcal{H}^{\sigma}(I)$, where $\sigma = \min\{\sigma_1, \sigma_2, \sigma_3\}$ and $\|S\|_{\sigma} = \sup_{k \in \mathbb{N}} \|\alpha_k\|_{\sigma}$. Then, we have*

$$1 \leq \dim_H \left(Graph(f_u^S)\right) \leq \underline{\dim}_B\left(Graph(f_u^S)\right) \leq \overline{\dim}_B\left(Graph(f_u^S)\right) \leq 2 - \sigma,$$

where $\sigma = \min\{\sigma_1, \sigma_2, \sigma_3\}$.

Proof It is known [17] that

$$1 \leq \dim_H \left(\mathrm{Graph}(f_u^S)\right) \leq \underline{\dim}_B\left(\mathrm{Graph}(f_u^S)\right) \leq \overline{\dim}_B\left(\mathrm{Graph}(f_u^S)\right).$$

From Remark 3.7, we say that $f, \alpha_{i,k}$ and u_k are elements of $\mathcal{H}^\sigma(I)$, where $\sigma = \min\{\sigma_1, \sigma_2, \sigma_3\}$. We define a sequence of mappings $T^{\alpha_k} : \mathcal{H}_f^\sigma(I) \to \mathcal{H}_f^\sigma(I)$ by

$$(T^{\alpha_k} g)(x) = f(x) + \alpha_{i,k}(Q_i(x))\ (g - u_k)(Q_i(x))$$

for all $x \in I_i$, where $i \in \mathbb{N}_N$. Then we have

$$\begin{aligned}
\left[(T^{\alpha_k} g)\right]_\sigma &= \max_{i\in\mathbb{N}_N} \sup_{x,y\in I_i,\ x\neq y} \frac{|T^{\alpha_k} g(x) - T^{\alpha_k} g(y)|}{|x-y|^\sigma} \\
&\leq \max_{i\in\mathbb{N}_N} \sup_{x,y\in I_i,\ x\neq y} \left[\frac{|f(x) - f(y)|}{|x-y|^\sigma} \right. \\
&\quad + \frac{|\alpha_{i,k}(x)|\left|(g-u_k)(Q_i(x)) - (g-u_k)(Q_i(y))\right|}{|x-y|^\sigma} \\
&\quad \left. + \frac{\left|(g-u_k)(Q_i(y))\right||\alpha_{i,k}(x) - \alpha_{i,k}(y)|}{|x-y|^\sigma} \right] \\
&\leq [f]_\sigma + \frac{\|S\|_\infty}{(\min\{|a_i|\})^\sigma}\left([g]_\sigma + [u_k]_\sigma\right) + \|g - u_k\|_\infty [S]_\sigma,
\end{aligned}$$

which shows that T^{α_k} is well-defined. Let $g, h \in \mathcal{H}_f^\sigma(I)$. We have

$$\begin{aligned}
\|T^{\alpha_k} g - T^{\alpha_k} h\|_\sigma &= \|T^{\alpha_k} g - T^{\alpha_k} h\|_\infty + [T^{\alpha_k} g - T^{\alpha_k} h]_\sigma \\
&\leq \|\alpha_k\|_\infty \|g - h\|_\infty + \frac{\|\alpha_k\|_\infty}{(\min\{|a_i|\})^\sigma}[g-h]_\sigma + [\alpha_k]_\sigma \|g-h\|_\infty \\
&\leq \max\left\{\|S\|_\sigma, \frac{\|S\|_\infty}{(\min\{|a_i|\})^\sigma}\right\}\|g - h\|_\sigma.
\end{aligned} \tag{3.6}$$

This yields that each T^{α_k} is a contraction mapping on $\mathcal{H}_f^\sigma(I)$. It is clear to observe that the sequence $\{\|T^{\alpha_k} g - g\|_\sigma\}$ is bounded. Using Proposition 3.2, the backward trajectories $\Psi_k := T^{\alpha_1} \circ T^{\alpha_2} \circ \cdots \circ T^{\alpha_k}$ of (T^{α_k}) converge for every $g \in \mathcal{H}_f^\sigma(I)$ to a unique attractor $f_u^S \in \mathcal{H}_f^\sigma(I)$. Since $f_u^S \in \mathcal{H}_f^\sigma(I)$, we obtain (Cf. [17, Corrollary 11.2] that $\overline{\dim}\left(\mathrm{Graph}(f_u^S)\right) \leq 2 - \sigma$. This proves the result. □

3.6 Some Approximations

Here, we return to the non-stationary α-fractal functions in the function space $C(I)$. First we recall the following definition of Schauder basis.

Definition 3.8 Let X be a Banach space. A sequence $\{x_n\}$ of X is a Schauder basis if for every $x \in X$, there exists a unique representation of x as $x = \sum\limits_{n=1}^{\infty} a_n x_n$, where $\{a_n\}$ is a sequence of scalars.

In Banach spaces, Schauder bases are important for applications in operator equations. We refer the reader to [26] for details of different Schauder bases of $C([0,1])$.

Theorem 3.8 *The space $C(I)$ admits a Schauder basis consisting of non-stationary α-fractal functions.*

Proof Let $\{f_m\}$ be a Schauder basis of $C(I)$ with the associated coefficient functionals $\{\lambda_m\}$. Consider the scaling sequence $\{\alpha^m = \{\alpha_k^m\}_{m\in\mathbb{N}}\}$. Suppose that α^m is chosen such that $\|\alpha\|_\infty < \frac{1}{1+K_L}$. Then, $\mathcal{F}_u^{S^m}$ is a topological automorphism for each $m \in \mathbb{N}$. Let $f \in C(I)$. Then $(\mathcal{F}_u^{S^m})^{-1}(f) \in C(I)$ and

$$(\mathcal{F}_u^{S^m})^{-1}(f) = \sum_{m=1}^{\infty} \lambda_m\left((\mathcal{F}_u^{S^m})^{-1}(f)\right) f_m.$$

Since $\mathcal{F}_u^{S^m}$ is a linear and continuous map,

$$f = \sum_{m=1}^{\infty} \lambda_m\left((\mathcal{F}_u^{B^m})^{-1}(f)\right) f_u^{B^m}.$$

To prove the uniqueness of the representation, let us assume another representation of f as

$$f = \sum_{m=1}^{\infty} \gamma_m f_u^{S^m}.$$

Using the continuity of $(\mathcal{F}_u^{S^m})^{-1}$, $(\mathcal{F}_u^{S^m})^{-1}(f) = \sum_{m=1}^{\infty} \gamma_m f_m$ and hence $\gamma_m = \lambda_m\left((\mathcal{F}_u^{S^m})^{-1}(f)\right)$, $m \in \mathbb{N}$. Thus, $\{f_u^{S^m}\}$ is a Schauder basis of $C(I)$ consisting of self-referential functions. □

Definition 3.9 Let $p \in C(I)$ be a polynomial. Using the definition of the fractal operator, we define the associated non-stationary α-fractal polynomial as $\mathcal{F}_u^S(p) = p_u^S$. For notational simplicity, we denote it by p^S. For stationary fractal polynomial see [22]. Let $\mathcal{P}(I) \subset C(I)$ be the space of all polynomials. We denote $\mathcal{P}^S(I)$ as the image space $\mathcal{F}_u^S(\mathcal{P}(I))$.

Notation Let $\mathcal{P}_m(I)$ be the set of all polynomials of degree at most m defined on I. That is,

$$\mathcal{P}_m(I) = \left\{p(x) = \sum_{i=0}^{m} a_i x^i : a_i \in \mathbb{R}, 0 \le i \le m\right\}.$$

We let $\mathcal{P}_m^\alpha(I) = \mathcal{F}_u^S(\mathcal{P}_m(I))$. □

Remark 3.8 Note that using the method of fractal perturbation, we can always obtain a new class of smooth/non-smooth approximant for any $X \subset C(I)$. Also, in real life applications, irregular approximant with a specified roughness (quantified in terms of the box counting dimension) is more advantageous than the classical smooth approximant. In all these cases, one can deal with the perturbed approximation class $\mathcal{F}_u^S(X)$.

Let us recall some basic results from approximation theory [10].

Definition 3.10 Let $(X, \|.\|)$ be a normed linear space over the field $\mathbb{K}$. For a nonempty set $V \subseteq X$, and $x \in X$, distance from x to V is defined as $d(x,V) = \inf\{\|x - v\| : v \in V\}$. v^* is a best approximant to x from V if there exists an element $v^*(x) \in V$ such that $\|x - v^*\| = d(x, V)$. We call A subset V of X is called proximinal (proximal or existence set) if for each $x \in X$ a best approximant $v^*(x) \in V$ of x exists.

Theorem 3.9 *Let V be a finite dimensional subspace of the normed linear space X. Then for each $x \in X$, there exists a best approximant from V.*

The following theorem is a direct consequence of Theorem 3.9, and the finite dimension of $\mathcal{P}_m^S(I)$.

Theorem 3.10 *Consider $(C(I), \|.\|_\infty)$. Then for each $f \in C(I)$, a best approximant p_f^S in $\mathcal{P}_m^S(I)$ exists.*

Theorem 3.11 *Let $f \in C(I)$, and $L_k : C(I) \to C(I)$, $L_k \neq Id$ be bounded linear operators satisfying $(L_k f)(x_i) = f(x_i) \quad \forall\ i \in \{0, N\},\ k \in \mathbb{N}$. For any $\epsilon > 0$, partition Δ of the interval I, there exists a fractal polynomial p^S such that*

$$\|f - p^S\|_\infty < \epsilon.$$

Proof Let $\epsilon > 0$ be given. Using the Stone-Weierstrass theorem, we obtain a polynomial p in two variables such that

$$\|f - p\|_\infty < \frac{\epsilon}{2}.$$

Fix a partition Δ of the interval I, bounded linear operators $L_k : C(I) \to C(I)$, $L_k \neq Id$ satisfying $(L_k f)(x_i) = f(x_i) \quad \forall\ i = 0, N,\ k \in \mathbb{N}$. Select $\alpha_{i,k} \in C(I)$ with $\|S\|_\infty = \sup\limits_{k \in \mathbb{N}} \|\alpha_k\|_\infty < 1$ such that

$$\|S\|_\infty < \frac{\frac{\epsilon}{2}}{\frac{\epsilon}{2} + K_L \|p\|_\infty},$$

where $K_L = \sup\limits_{k \in \mathbb{N}} \|Id - L_k\|$. Then we have

$$\begin{aligned}
\|f - p^S\|_\infty &\leq \|f - p\|_\infty + \|p - p^S\|_\infty \\
&\leq \|f - p\|_\infty + \frac{\|S\|_\infty}{1 - \|S\|_\infty} K_L\ \|p\|_\infty \\
&< \frac{\epsilon}{2} + \frac{\epsilon}{2} \\
&= \epsilon.
\end{aligned}$$

In the above, the first inequality is just the triangle inequality, second follows from Theorem 3.5 and third is obvious. □

Remark 3.9 Note that, in the above proof, we considered $\alpha_{i,k} \in C(I)$ with $\|S\|_\infty < \frac{\frac{\epsilon}{2}}{\frac{\epsilon}{2} + K_L\ \|p\|_\infty}$. One can observe that, in some cases, the sequence $\alpha_{i,k}$ may converge to 0, and thus p^S may lose self-referentiality. Consequently, it may lead to behave as a traditional polynomial. On the other hand, one can select arbitrary $\alpha_{i,k} \in C(I)$ with $\|S\|_\infty < 1$, and choose bounded linear operators $L_k : C(I) \to C(I)$, $L_k \neq Id$ with $(L_k f)(x_i) = f(x_i), \quad \forall\ i = 0, N$ such that

$$K_L < \frac{1 - \|S\|_\infty}{\|S\|_\infty \|p\|_\infty} \frac{\epsilon}{2}.$$

For this case, we believe that the graph of the corresponding fractal polynomial p^S has box dimension ≥ 2. In particular, it shows fractality character which differs from the traditional polynomial.

The following result is a direct consequence of the above theorem.

Theorem 3.12 *The set of non-stationary α-fractal polynomials with non-null scale vector is dense in $C(I)$.*

In the next theorem, we provide the denseness of a class of non-stationary α-fractal polynomials which is a proper subset of the dense set considered above.

Theorem 3.13 *If $\|S\|_\infty < \frac{1}{1+K_L}$, then $\mathcal{P}^\alpha(I)$ is dense in $C(I)$.*

Proof $\mathcal{F}_u^S$ is a topological automorphism under the give assumption on α. Let $f \in C(I)$. Then from the Stone-Weierstrass theorem, we obtain a sequence of polynomials (p_n) with $p_n \to (\mathcal{F}_u^S)^{-1}(f)$ in the uniform norm. Also, as $\mathcal{F}_u^S$ is bounded, we have $p_n^S := \mathcal{F}_u^S(p_n) \to f$ as $n \to \infty$. □

Let us assume $I = [-1, 1]$ for the following result.

Definition 3.11 [12] Consider $(C(I), \|.\|_\infty)$ and $f \in C(I)$. The Ditzian-Totik modulus of smoothness is defined as

$$\omega_r^\phi(f, \delta) = \sup_{0<h\leq\delta} \left|\overline{\Delta}_{h\phi(x)}^r f(x)\right|,$$

where $\phi(x) = \sqrt{1-x^2}$ and r-th symmetric difference of the function f is given by

$$\overline{\Delta}_{h\phi(x)}^r f(x) = \sum_{k=0}^{r}(-1)^k\binom{r}{k} f\left(x + h\phi(x)(\frac{r}{2} - k)\right)$$

if $\left(x \pm rh\phi(x)/2\right) \in [-1, 1]$, $\overline{\Delta}_{h\phi(x)}^r f(x) = 0$ elsewhere.

Theorem 3.14 *[12]If $f \in C(I)$, a sequence of polynomials $(P_m(f))_{m\in\mathbb{N}}$ exists, with degree $\leq m$, such that*

$$\|f - P_m(f)\| \leq C\ \omega_2^\phi\left(f; \frac{1}{m}\right),$$

where $C > 0$ is independent of f, m and $\omega_2^\phi\left(f; \frac{1}{m}\right)$ is the Ditzian-Totik modulus of smoothness with $\phi(x) = \sqrt{1-x^2}$.

Notation Let us define $E_m(f) := \inf\{\|f-p\|_\infty : p \in \mathcal{P}_m(I)\}$ and $E_m^S(f) := \inf\{\|f-p^S\|_\infty : p^S \in \mathcal{P}_m^S(I)\}$.□

Theorem 3.15 *Let $f \in C(I)$, then*

$$E_m^S(f) \leq C\,\frac{1 + \|S\|_\infty(K_L - 1)}{1 - \|S\|_\infty}\omega_2^\phi\left(f; \frac{1}{m}\right) + \frac{\|S\|_\infty K_L}{1 - \|S\|_\infty}\|f\|_\infty,$$

where $C > 0$ is an absolute constant and $K_L = \sup_{k\in\mathbb{N}} \|Id - L_k\|$.

Proof Let $f \in C(I, \mathbb{R})$. Let $p_f \in \mathcal{P}_m(I)$ be a best approximant to f. That is, $E_m(f) = \|f - p_f\|_\infty$. Using the previous theorem, we estimate a bound for $E_m^S(f)$ in the following manner:

$$\begin{aligned}
E_m^S(f) &\le \|f-(p_f)^S\|_\infty \\
&\le \|f-p_f\|_\infty + \|p_f-(p_f)^S\|_\infty \\
&\le E_m(f) + \frac{\|S\|_\infty K_L}{1-\|S\|_\infty}\|p_f\|_\infty \\
&\le E_m(f) + \frac{\|S\|_\infty K_L}{1-\|S\|_\infty}\|p_f-f+f\|_\infty \\
&\le E_m(f)\Big[1+\frac{\|S\|_\infty K_L}{1-\|S\|_\infty}\Big] + \frac{\|S\|_\infty K_L}{1-\|S\|_\infty}\|f\|_\infty \\
&\le \frac{1+\|S\|_\infty (K_L-1)}{1-\|S\|_\infty}E_m(f) + \frac{\|S\|_\infty K_L}{1-\|S\|_\infty}\|f\|_\infty.
\end{aligned}$$

The proof of the theorem follows now using Theorem 3.14. □

3.7 Concluding Remarks

In this chapter, we have discussed different aspects of non-stationary univariate α-fractal functions. At first, we have introduced the idea of non-stationary univariate α-fractal functions. Here we have used a sequence of maps for the non-stationary iterated function systems. For a suitable selection of the IFS parameters, we defined the non-stationary α-fractal operator and discussed some of its properties. Then we showed the existence of a non-trivial closed invariant subspace for the associated fractal operator. Also, we have estimated the bound of the box and Hausdorff dimension of the proposed interpolant. In the end, we discussed some approximation properties of the non-stationary α-fractal functions in $C(I)$.

References

[1] Agrawal, V., T. Som and S. Verma. 2021. On bivariate fractal approximation. arXiv preprint arXiv:2101.07146.

[2] Akhtar, Md. N., M.G.P. Prasad and M.A. Navascuès. 2017. Box dimension of α-fractal function with variable scaling factors in subintervals. Chaos, Solitons and Fractals, 103: 440–449.

[3] Barnsley, M.F. 1986. Fractal functions and interpolation. Constr. Approx., 2: 303–32.

[4] Barnsley, M.F. and P.R. Massopust. 2015. Bilinear fractal interpolation and box dimension. J. Approx. Theory, 192: 362–378.

[5] Barnsley, M.F., J.E. Hutchinson and Ö. Stenflo, 2008. V-variable fractals: Fractals with partial self similarity. Adv. Math., 218(6): 2051–2088.

[6] Bollobás, B. 1999. Linear Analysis, and Introductory Course. Cambridge University Press, 2nd ed.

[7] Chand, A.K.B., S. Jha and M.A. Navascués. 2020. Kantorovich-Bernstein α-fractal functions in L^p spaces. Quaest. Math., 43(2): 227–241.

[8] Chand, A.K.B. and G.P. Kapoor. 2006. Generalized cubic spline fractal interpolation functions. SIAM J. Numer. Anal., 44: 655–676.

[9] Chandra, S. and S. Abbas. 2021. The calculus of bivariate fractal interpolation surfaces. Fractals, 29(3): 2150066.

[10] Cheney, E.W. 1982. Introduction to Approximation Theory. 2nd ed. New York: Chelsea.

[11] Dyn, N., D. Levin and P. Massopust. 2020. Attractors of trees of maps and of sequences of maps between spaces and applications to subdivision. J. Fixed Point Theory Appl., 22(1).
[12] Gal, S.G. 2008. Shape-preserving Approximation by Real and Complex Polynomials. Birkhäuser Boston, Inc., Boston.
[13] Hutchinson, J. 1981. Fractals and self similarity. Indiana Uni. Math. J., 30(5): 713–747.
[14] Jha, S., A.K.B. Chand and M.A. Navascués. 2021. Approximation by shape preserving fractal functions with variable scalings. Calcolo, 58(1): 24.
[15] Jha, S. and S. Verma. 2021. Dimensional analysis of α-fractal functions. Results in Mathematics, 76(4): 1–24.
[16] Levin, D., N. Dyn and P. Viswanathan. 2019. Non-stationary versions of fixed-point theory, with applications to fractals and subdivision. J. Fixed Point Theory Appl., 21: 1–25.
[17] Falconer, K.J. 1999. Fractal Geometry: Mathematical Foundations and Applications. John Wiley Sons Inc., New York.
[18] Limaye, B.V. 2016. Linear Functional Analysis for Scientists and Engineers. Springer, Singapore.
[19] Luor, D.-C. 2018. Fractal interpolation functions with partial self similarity. J. Math. Anal. Appl., 464: 911–923.
[20] Massopust, P.R. 2016. Fractal Functions, Fractal Surfaces, and Wavelets. 2nd ed., Academic Press, San Diego.
[21] Massopust, P.R. 2019. Non-stationary fractal interpolation. Mathematics, 7(8): 666.
[22] Navascués, M.A. 2005. Fractal polynomial interpolation. Z. Anal. Anwend., 25(2): 401–418.
[23] Navascués, M.A. 2010. Fractal approximation. Complex Anal. Oper. Theory, 4(4): 953–974.
[24] Navascués, M.A. and S. Verma. 2021. Non-stationary $\alpha-$fractal surfaces. Preprint.
[25] Ri, S. 2017. A new nonlinear fractal interpolation function. Fractals, 25(6).
[26] Semadeni, Z. 1982. Schauder bases in Banach spaces of continuous functions. Lecture Notes in Mathematics, 918, Springer-Verlag, Berlin.
[27] Verma, S. and P. Viswanathan. 2019. A revisit to α-fractal function and box dimension of its graph. Fractals, 27(6): 1950090.
[28] Verma, S. and P. Viswanathan. 2020. A fractal operator associated with bivariate fractal interpolation functions on rectangular grids. Results Math, 75: 28.
[29] Verma, S. and P. Viswanathan. 2020. A fractalization of rational trigonometric functions. Mediterranean Journal of Mathematics, 17: 1–23.
[30] Verma, S. and P.R. Massopust. 2020. Dimension preserving approximation. arXiv preprint arXiv:2002.05061.
[31] Vijender, N. 2018. Fractal perturbation of shaped functions: convergence independent of scaling. Mediterr. J. Math, 15(6): 1–16.
[32] Vijender, N. 2019. Bernstein fractal trigonometric approximation. Acta Appl. Math., 159: 11–27.

Chapter 4

Fractal Calculus

Alireza Khalili Golmankhaneh,[1,*] Kerri Welch,[2] TMC Priyanka[3] and A Gowrisankar[3]

4.1 Introduction

Fractal geometry is introduced by Benoit Mandelbrot as an extension of Euclidean geometry. The sets considered in fractal geometry are more irregular than the sets studied in Euclidean geometry. Fractal patterns are extremely familiar because nature is full of fractals. For instance: branches of trees, snowflakes, animal circulatory systems, plants and leaves, lightning and electricity, clouds, crystals, rivers, coastlines, mountains, seashells, hurricanes are all examples of fractal patterns. Self-similarity is the most fruitful concept of fractals. In recent times, the theory of fractals has been applied to many natural phenomena due to their self-similarity and irregularity of nature. Based on the self-similar property, fractals are classified into two types namely random fractals (statistically self-similar) and deterministic fractals (exactly self-similar). Mountain ranges, coastlines, and lightening are some of the examples of random fractals seen in nature whereas the Sierpinski triangle, and Cantor set are the mathematical examples of deterministic fractals.

Fractals are described as geometrical shapes such that their fractal dimension is greater than their topological dimension [1–13]. Analysis on fractals has an important role in the applications and modeling of processes with fractal structures. Hence, fractal analysis was formulated using different methods like harmonic analysis, the probabilistic method, fractional spaces, fractional calculus, non-standard analysis and, time scale method [14–21, 23–28, 28–33]. Most of the complex physical structures found in the real world can be modeled as fractals. As the fractals are often non-differentiable structures, the techniques of ordinary calculus cannot be applied to them. Consequently, the ordinary calculus is not sufficient to handle the continuous and non-differentiable fractal curves like the trajectory of a quantum particle, structure of polymer molecules. Though there are a variety of approaches to this problem, many researchers have established the importance of employing fractional calculus to explore and study the characteristics of fractal curves.

Non-conservative systems in mechanics can be modelled using fractional derivatives having nonlocality properties. However, choosing a good fractional derivative for practical situations is not always a straightforward task, because there are so many different definitions of fractional derivatives, and new ones appear every day. Besides, as fractional derivatives are non-local operators, they are not always appropriate for dealing with the local scaling behavior of fractal functions. Therefore, the fractional calculus has been renormalized as fractal calculus to provide local fractional operators. In [33], the authors have introduced a special calculus on fractals $F \subset \mathbb{R}$ called F^{α}-calculus as a simple and direct approach. For the dimension α of the fractal F, it defines derivatives and integrals of order $\alpha \in (0, 1]$ on fractals. These derivatives and integrals are called as F^{α}-derivatives and F^{α}-integrals respectively. The definition of the derivative as a limit of a quotient and the

[1] Department of Physics, Urmia Branch, Islamic Azad University, Urmia, Iran.
[2] Faculty at California Institute of Integral Studies, San Francisco, CA, USA.
[3] Department of Mathematics, School of Advanced Sciences, Vellore Institute of Technology, Vellore 632 014, Tamil Nadu, India.
Emails: kwelch@ciis.edu; priyankamohan195@gmail.com; gowrisankargri@gmail.com
[*] Corresponding author: alirezakhalili2002@yahoo.co.in

Riemann-like technique to define integral are algorithmically favorable and not dependent on the underlying fractal.

The fractal calculus called F^{α} calculus can be viewed as a generalization of classical calculus since it contains several results that are analogues to classical calculus. F^{α} calculus has been used as an effective tool for modeling processes on fractals. In [41], the authors have introduced the concept of non-local fractal calculus. The comparison between the local and non-local linear differential equations is presented in [42]. Besides, for the non-local fractal differential equations, some applications related to physics are also discussed in [42]. Fractal derivatives and integrals of functions are defined on the Cantor Tarton spaces in [37]. Fractal integral and fractal derivative for the fractal curve called von Koch curve is investigated in [36]. In the seminal papers [34, 35], Riemann-like calculus called F^{α}-calculus (F^{α}-C) has been formulated to include the derivatives and integrals of functions with support from totally disconnected sets such as Cantor sets and non-differentiable curves like Koch and Cesáro curves [36]. To continue this study, fractal calculus (=local fractal calculus) is generalized to involve derivatives and integrals of functions with the support of the Cantor cubes and the Cantor tartan spaces [37–39]. In view of the ordinary fractional calculus with the non-local properties, non-local fractal calculus has been defined to model the processes with memory on fractal spaces [40, 42].

Reasons why fractional calculus is not suitable for fractals and physical processes:

1. The order of the fractional derivatives is non-integer and it does not have a geometrical or physical meaning.
2. The fractional derivatives involve real-line measures like length, whereas a fractal has a different measure for example, the Hausdorff Measure.
3. The fractional diffusion equation provides a mathematical model for sup-and super diffusion but violates simple forms of the central limit theorem.

The main difference between the fractional calculus and fractal calculus is that their operators are non-local and local respectively. Fractal calculus in both local and non-local form provides us mathematical framework to express physical law in spaces with integer and fractional dimensions. Local fractal calculus is a simple, constructive, and algorithmic approach to perform analysis on fractals. The advantages include the following properties:

1. In physics, the local fractal derivative is crucial for two reasons: first, it ensures that causality is not violated, and second, measurement in physics is local.
2. The order of the fractal derivative is non-integer and has geometrical meaning, being equal to the dimension of support of the function.
3. The order of the fractal derivative also has physical meaning by having a relationship with the spectral dimension.

Cantor set is one of the interesting fractal structures of study in fractal analysis, it is constructed by repeatedly removing the middle open third interval of [0,1]. Several researchers have investigated various aspects of Cantor set. In this chapter, local fractal calculus on thin Cantor set is explored. Further, non-local fractal derivatives like fractal Riemann-Liouville derivative and fractal Caputo derivative of the Cantor set are shortly narrated. Besides, applications of fractal calculus in some scientific domains are also discussed to help realize its practical value.

The organization of this chapter is as follows: In Section 2, the construction of Cantor set and its local fractal calculus are investigated. Section 3 briefly discusses some formulae of non-local fractal calculus. The Fourier transform and local fractal Fourier transform of different functions are presented in Section 4. Similarly, Laplace transform and local fractal Laplace transform of various functions are illustrated in Section

5. Section 6 precisely discusses the scaling properties of Cantor set. Section 7 explores the applications of fractal calculus on classical mechanics, quantum mechanics and optics. Finally, a conclusion has been provided in Section 8.

4.2 Local Fractal Calculus

In this section, we have summarized fractal calculus on different fractals. We present the stages which set up **the thin Cantor set**(= middle-ϵ Cantor set) (See [43] for more details).

Let us consider an interval, $\mathbf{I} = [a, b]$, then remove an open interval of fraction (=$0 < \epsilon < 1$) of length from the middle of **I**. By carrying out the same procedure, we form a thin Cantor set as follows:
Step 1.

$$F_1^\epsilon = [0, \frac{1}{2}(1-\epsilon)] \cup [\frac{1}{2}(1+\epsilon), 1], \tag{4.1}$$

Step 2.

$$F_2^\epsilon = [0, \frac{1}{4}(1-\epsilon)^2] \cup [\frac{1}{4}(1-\epsilon^2), \frac{1}{2}(1-\epsilon)] \\ \cup [\frac{1}{2}(1+\epsilon) + \frac{1}{2}((1+\epsilon), \frac{1}{2}(1-\epsilon)^2)] \cup [\frac{1}{2}(1+\epsilon)(1+\frac{1}{2}(1-\epsilon)), 1] \tag{4.2}$$

$$\vdots$$

Finally, we arrive at after m steps

$$F^\epsilon = \bigcap_{m=1}^{\infty} F_m^\epsilon, \tag{4.3}$$

which is called middle-ϵ Cantor set/thin Cantor-like set.
The Hausdorff dimension of the thin Cantor set is defined by [43]

$$Dim_H(F^\epsilon) = \frac{\log(2)}{\log(2) - \log(1-\epsilon)}. \tag{4.4}$$

In Figure 4.2(a), we show the process that establish the middle-ϵ Cantor set. The flag function is defined by [34, 35],

$$\Theta(F^\epsilon, \mathbf{I}) = \begin{cases} 1 & \text{if } F^\epsilon \cap \mathbf{J} \neq \emptyset \\ 0 & \text{otherwise.} \end{cases} \tag{4.5}$$

Let $\mathbf{Q}_{[a,b]} = \{a = t_0, t_1, t_2, \ldots, t_n = b\}$ be a subdivision of **I**. Then, $\Upsilon^\alpha[F^\epsilon, \mathbf{Q}]$ is defined in [34, 35] by

$$\Upsilon^\alpha[F^\epsilon, \mathbf{Q}] = \sum_{i=1}^{n} \Gamma(\alpha+1)(t_i - t_{i-1})^\alpha \Lambda(F^\epsilon, [t_{i-1}, t_i]), \tag{4.6}$$

where $0 < \alpha \leq 1$.

The mass function $\mathcal{M}^{\alpha}(F^{\epsilon}, a, b)$ is given in [34, 35] by

$$\begin{aligned}\mathcal{M}^{\alpha}(F^{\epsilon}, a, b) &= \lim_{\delta\to 0} \mathcal{M}^{\alpha}_{\delta}(F^{\epsilon}, a, b)\\ &= \lim_{\delta\to 0}\left(\inf_{\mathbf{Q}_{[a,b]}:|Q|\leq\delta} \Upsilon^{\alpha}[F^{\epsilon}, \mathbf{Q}]\right),\end{aligned} \tag{4.7}$$

here, by taking infimum over all subdivisions $\mathbf{Q}$ of $[a, b]$ satisfying $|\mathbf{Q}| := \max_{1\leq i\leq n}(t_i - t_{i-1}) \leq \delta$.
In [34, 35], the integral staircase function of the fractal sets is defined as

$$S^{\alpha}_{F^{\epsilon}}(t) = \begin{cases}\mathcal{M}^{\alpha}(F^{\epsilon}, t_0, t) & \text{if} \quad t \geq t_0\\ -\mathcal{M}^{\alpha}(F^{\epsilon}, t_0, t) & \text{otherwise,}\end{cases} \tag{4.8}$$

where t_0 is an arbitrary real and fixed number.
In Figure 4.2(a), we plot Eq. (4.8) middle-ϵ Cantor set by letting $\epsilon = 1/2$.
The γ-dimension of $F^{\epsilon} \cap [a, b]$ is

$$\begin{aligned}\dim_{\gamma}(F^{\epsilon} \cap [a, b]) &= \inf\{\alpha : \mathcal{M}^{\alpha}(F^{\epsilon}, a, b) = 0\}\\ &= \sup\{\alpha : \mathcal{M}^{\alpha}(F^{\epsilon}, a, b) = \infty\}.\end{aligned} \tag{4.9}$$

In Figure 4.2(c), we have obtained γ-dimension in view of Eq. (4.9).
For a middle-ϵ Cantor set fractal, the characteristic function is defined by

$$\chi_{F^{\epsilon}}(\alpha, t) = \begin{cases}\frac{1}{\Gamma(\alpha+1)}, & t \in F^{\epsilon};\\ 0, & \text{otherwise}.\end{cases} \tag{4.10}$$

In Figure 4.2, we have plotted characteristic function for the middle-ϵ Cantor set by choosing $\epsilon = 1/2$.
F$^{\alpha}$-limit (=**fractal limit**) of $h : F^{\epsilon} \to \text{Re}$ is given by

$$\text{and} \quad |z - t| < \delta \Rightarrow |h(z) - l| < \epsilon, \quad \forall t, z \in F^{\epsilon} \tag{4.11}$$

if l exists, namely

$$l = \text{F}^{\alpha}\text{-}\lim_{z\to t} h(t). \tag{4.12}$$

F$^{\alpha}$-continuity (=**fractal continuity**) of h is defined by

$$h(z) = \text{F}^{\alpha}\text{-}\lim_{z\to t} h(t). \tag{4.13}$$

F$^{\alpha}$-differentiation (=**fractal derivative**) of h on α-perfect set, is defined by [34, 35],

$$D^{\alpha}_{F^{\epsilon}} h(t) = \begin{cases}\text{F}^{\alpha}\text{-}\lim_{z\to t} \frac{h(z)-h(t)}{S^{\alpha}_{F^{\epsilon}}(z)-S^{\alpha}_{F^{\epsilon}}(t)}, & \text{if } z \in F^{\epsilon},\\ 0, & \text{otherwise.}\end{cases} \tag{4.14}$$

F^{α}-integral (=**fractal integral**) of h on $[a, b]$ is denoted by $\int_a^b h(t)d^{\alpha}_{F^{\epsilon}}t$ and is approximately given by [34, 35]

$$\int_a^b h(t)d^{\alpha}_{F^{\epsilon}}t \approx \sum_{i=1}^{n} h_i(t)(S^{\alpha}_{F^{\epsilon}}(t_j) - S^{\alpha}_{F^{\epsilon}}(t_{j-1})). \tag{4.15}$$

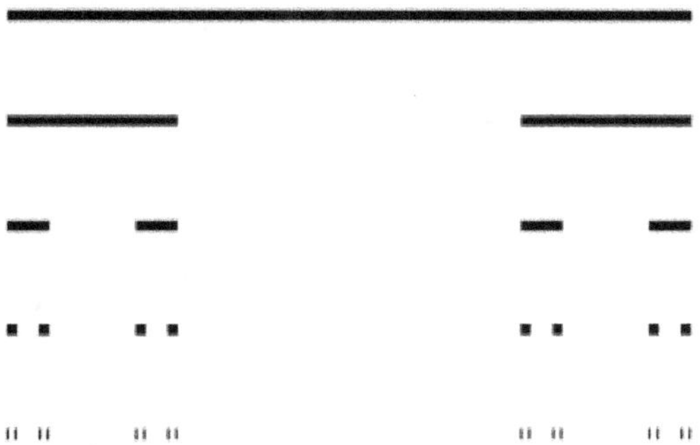

(a) The thin Cantor-like set F^{ϵ} ($\epsilon = 1/3$) by iteration.

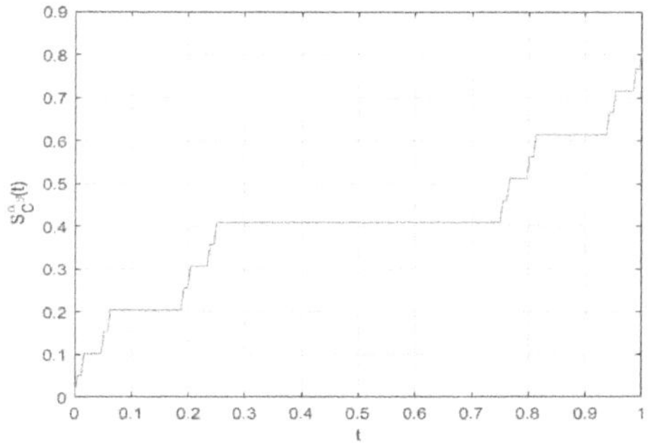

(b) The integral staircase function for the thin Cantor set F^{ϵ} for the case of ($\epsilon = 1/2$).

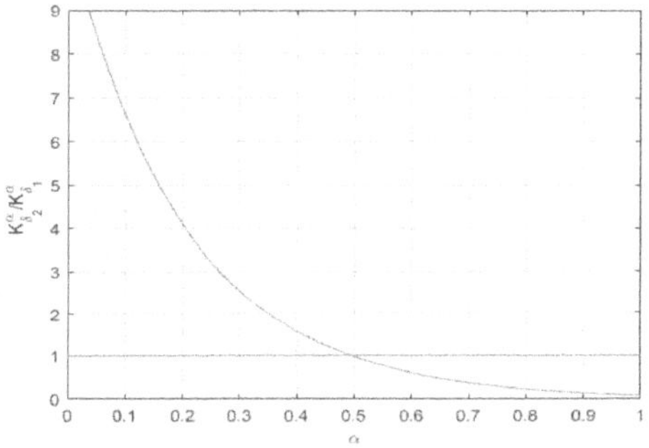

(c) Ψ – dimension of the thin Cantor set F^{ϵ} ($\epsilon = 1/2$).

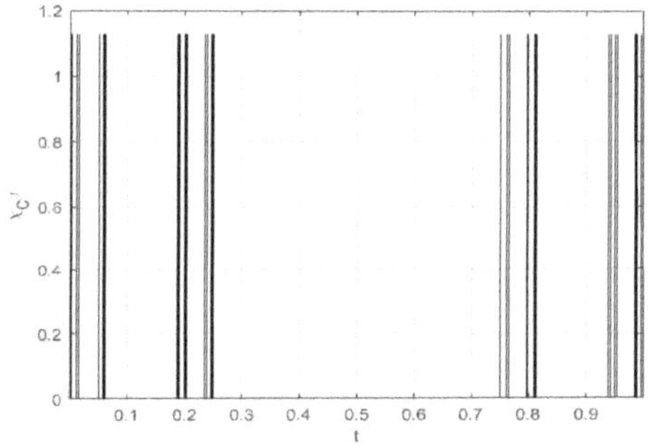

(d) Characteristic function of thin Cantor set F^{ϵ} with ($\epsilon = 1/2$).

Fig. 4.1 Graphs corresponding to thin Cantor set F^{ϵ} with ($\epsilon = 1/2$).

4.3 Non-local Fractal Calculus

In this section, we review the non-local fractal derivatives [42].

The fractal right-sided Riemann-Liouville integral of order $\nu \in \text{Re}$ is defined by:

$$
{}_x\mathcal{I}_b^{\nu} h(x) := \frac{1}{\Gamma_{F^{\epsilon}}^{\alpha}(\nu)} \int_x^b \frac{h(t)}{(S_{F^{\epsilon}}^{\alpha}(x) - S_{F^{\epsilon}}^{\alpha}(t))^{\alpha-\nu}} d_{F^{\epsilon}}^{\alpha} t, \tag{4.16}
$$

where $S_{C^{\alpha}}^{\alpha}(b) > S_{C^{\alpha}}^{\alpha}(x)$ and $\nu \neq \alpha$.

The fractal right-sided Riemann-Liouville derivative of order $\nu \in \text{Re}$ is defined by:

$$
{}_x\mathcal{D}_b^{\nu} h(x) := \frac{1}{\Gamma_{F^{\epsilon}}^{\alpha}(n-\nu)} (D_{F^{\epsilon}}^{\alpha})^n \int_x^b \frac{h(t)}{(S_{F^{\epsilon}}^{\alpha}(x) - S_{F^{\epsilon}}^{\alpha}(t))^{-n\alpha+\nu+\alpha}} d_{F^{\epsilon}}^{\alpha} t, \tag{4.17}
$$

where $h(x) \in F^{\alpha n}[a, b]$ (=n times α-differentiable), and $n\alpha - \alpha \leq \nu < \alpha n$.

The fractal right-sided Caputo derivative of order ν is defined by:

$$ {}_{x}^{C}\mathcal{D}_{b}^{\nu}h(x) := \frac{1}{\Gamma_{F^{\epsilon}}^{\alpha}(n-\nu)} \int_{x}^{b} \frac{(D_{F^{\epsilon}}^{\alpha})^{n}h(t)}{(S_{F^{\epsilon}}^{\alpha}(x) - S_{F^{\epsilon}}^{\alpha}(t))^{-n\alpha+\nu+\alpha}} d_{F^{\epsilon}}^{\alpha}t. \tag{4.18} $$

The fractal left-sided non-local derivatives can be defined in a similar way [42].
Some important formulae in fractal calculus:

$$ \begin{aligned}
& D_{F^{\epsilon}}^{\alpha} a\chi_{F^{\epsilon}} = 0, \quad a \text{ is constant}, \\
& D_{F^{\epsilon}}^{\alpha} S_{F^{\epsilon}}^{\alpha}(t) = \chi_{F^{\epsilon}}(\alpha, t), \\
& D_{F^{\epsilon}}^{\alpha} S_{F^{\epsilon}}^{\alpha}(t)^{m} = m S_{F^{\epsilon}}^{\alpha}(t)^{m-1}, \\
& D_{F^{\epsilon}}^{\alpha} \cos(S_{F^{\epsilon}}^{\alpha}(t)) = -\chi_{F^{\epsilon}}(\alpha, t)\sin(S_{F^{\epsilon}}^{\alpha}(t)), \\
& D_{F^{\epsilon}}^{\alpha}(f(t)g(t)) = D_{F^{\epsilon}}^{\alpha}(f(t))g(t) + f(t)D_{F^{\epsilon}}^{\alpha}(g(t)), \\
& {}_{0}\mathcal{D}_{t}^{\nu} a\chi_{F^{\epsilon}} = \frac{aS_{F^{\epsilon}}^{\alpha}(t)^{-\nu}}{\Gamma_{F^{\epsilon}}^{\alpha}(1-\nu)}, \\
& {}_{0}\mathcal{D}_{t}^{\nu} f(t)g(t) = \sum_{n=0}^{\infty} \binom{\nu}{n} {}_{0}\mathcal{D}_{t}^{n} f(t) {}_{0}\mathcal{D}_{t}^{\nu-n} g(t), \\
& \int S_{F^{\epsilon}}^{\alpha}(t)^{n} d_{F^{\epsilon}}^{\alpha}t = \frac{S_{F^{\epsilon}}^{\alpha}(t)^{n+1}}{n+1} + c, \quad c \text{ is constant}, \\
& f(t) = \sum_{n=1}^{\infty} \frac{D_{F^{\epsilon}}^{n\alpha} f(t)|_{t=a}}{n!} (S_{F^{\epsilon}}^{\alpha}(t) - S_{F^{\epsilon}}^{\alpha}(a)), \\
& {}_{x}\mathcal{I}_{b}^{\nu}\, {}_{x}\mathcal{D}_{b}^{\nu} h(x) = h(x) - \sum_{j=1}^{n} \frac{({}_{x}\mathcal{D}_{b}^{\nu-j} h(x))|_{(S_{C^{a}}^{\alpha}(b))}}{\Gamma(\nu+\alpha-j)} (S_{F^{\epsilon}}^{\alpha}(b) - S_{F^{\epsilon}}^{\alpha}(x))^{\nu-j}, \\
& {}_{a}\mathcal{I}_{x}^{\nu}\, {}_{a}^{C}\mathcal{D}_{x}^{\nu} h(x) = h(x) - \sum_{j=1}^{n} \frac{((D_{F^{\epsilon}}^{\alpha})^{j} h(x))|_{(S_{F^{\epsilon}}^{\alpha}(a))}}{\Gamma(j+\alpha)} (S_{F^{\epsilon}}^{\alpha}(x) - S_{F^{\epsilon}}^{\alpha}(a))^{j}, \\
& {}_{x}\mathcal{I}_{b}^{\nu}\, {}_{x}^{C}\mathcal{D}_{b}^{\nu} h(x) = h(x) - \sum_{j=1}^{n} \frac{((D_{F^{\epsilon}}^{\alpha})^{j} h(x))|_{(S_{F}^{\alpha}(b))}}{\Gamma(j+\alpha)} (S_{F^{\epsilon}}^{\alpha}(b) - S_{F^{\epsilon}}^{\alpha}(x))^{j}.
\end{aligned} \tag{4.19} $$

4.4 The Local Fractal Fourier Transform

The local fractal Fourier transform is defined by:

$$ \mathcal{F}\{f(t)\} = \mathbf{F}(w) = \frac{1}{\sqrt{2\pi}} \int_{-\infty}^{\infty} \exp\left(iS_{F^{\epsilon}}^{\alpha}(w)\, S_{F^{\epsilon}}^{\alpha}(t)\right) f(t)\, d_{F^{\epsilon}}^{\alpha}t. \tag{4.20} $$

The inverse local fractal Fourier transform of $\mathbf{F}(w)$ is defined by:

$$ \mathcal{F}^{-1}\{\mathbf{F}(w)\} = f(t) = \frac{1}{\sqrt{2\pi}} \int_{-\infty}^{\infty} \exp\left(-iS_{F^{\epsilon}}^{\alpha}(w)\, S_{F^{\epsilon}}^{\alpha}(t)\right) \mathbf{F}(w)\, d_{F^{\epsilon}}^{\alpha}w. \tag{4.21} $$

In Table 4.1, we have summarized the local fractal Fourier transform and Fourier transform of different functions.

Table 4.1 The local fractal Fourier transform and the Fourier transform.

Function	Fourier transform	Local fractal Fourier transform
$f(t)$	$f(w)$	$\mathbf{F}(w)$
$f(t)=\begin{cases}1, & \lvert t\rvert<b\\ 0, & \lvert t\rvert>b\end{cases}$	$\frac{2\sin w}{w}$	$\frac{2\sin S_{F\epsilon}^{\alpha}(w)}{S_{F\epsilon}^{\alpha}(w)}$
$\frac{1}{t^2+b^2}$	$\frac{\pi\exp(-bw)}{b}$	$\frac{\pi\exp\left(-bS_{F\epsilon}^{\alpha}(w)\right)}{b}$
$\frac{t}{t^2+b^2}$	$-i\pi\exp(bw)$	$-i\pi\exp\left(bS_{F\epsilon}^{\alpha}(w)\right)$
$f^{(n)}(t)$	$(iw)^n f(w)$	$\left(-iS_{F\epsilon}^{\alpha}(w)\right)^n\mathbf{F}(w)$
$t^n f(t)$	$i^n\frac{d^n f}{dw^n}$	$i^n D_{F\epsilon,w}^{n\alpha}(\mathbf{F}(w))$
$f(bt)\exp(ixt)$	$\frac{1}{b}f\left(\frac{w-x}{b}\right)$	$\frac{1}{b}\mathbf{F}\left(\frac{S_{F\epsilon}^{\alpha}(w)-S_{F\epsilon}^{\alpha}(x)}{b}\right)$
$\frac{df}{dx}$	$2\pi iwf(w)$	$2\pi iS_{F\epsilon}^{\alpha}(w)\mathbf{F}(w)$
$\int_{-\infty}^{x}f(t)dt$	$\frac{1}{iw}f(iw)$	$\frac{1}{iS_{F\epsilon}^{\alpha}(w)}\mathbf{F}(iw)$
$sgn(t)$	$\frac{2}{iw}$	$\frac{2}{iS_{F\epsilon}^{\alpha}(w)}$

4.5 The Local Fractal Laplace Transform

The local fractal Laplace transform is defined by:

$$B(w)=\mathcal{L}(f)=\int_{0}^{\infty}\exp\left(-S_{F\epsilon}^{\alpha}(t)\,S_{F\epsilon}^{\alpha}(w)\right)f(t)\,d_{F\epsilon}^{\alpha}t, \tag{4.22}$$

The inverse local fractal Laplace transform is given by

$$f(t)=\mathcal{L}^{-1}(B(w)). \tag{4.23}$$

Table 4.2 The fractal Laplace transform and the Laplace transform.

Function	Laplace transform	Local fractal Laplace transform
$f(t)$	$b(w)$	$B(w)$
c	$\frac{c}{w}$	$\frac{c}{S_{F^\epsilon}^{\alpha}(w)}$
$\exp(ct)$	$\frac{1}{w-c}$	$\frac{1}{S_{F^\epsilon}^{\alpha}(w)-c}$
$\sin(ct)$	$\frac{1}{w^2+c^2}$	$\frac{1}{S_{F^\epsilon}^{\alpha}(w)^2+c^2}$
t^m	$\frac{m!}{w^{m+1}}$	$\frac{m!}{S_{F^\epsilon}^{\alpha}(w)^{m+1}}$
$\frac{df}{dt}$	$wb(w)-f(0)$	$S_{F^\epsilon}^{\alpha}(w)B(w)-f(0)$
$\frac{d^2f}{dt^2}$	$w^2b(w)-wf(0)-\frac{df}{dt}\vert_0$	$S_{F^\epsilon}^{\alpha}(w)^2B(w)-S_{F^\epsilon}^{\alpha}(w)f(0)-D_{F^\epsilon}^{\alpha}f(t)\vert_0$
$\int_0^t f(t')dt'$	$\frac{b(w)}{w}$	$\frac{B(w)}{S_{F^\epsilon}^{\alpha}(w)}$

4.6 Discrete Scale Invariance

The discrete scale invariance gives log-periodic corrections to scaling and fractal structure. Here, we express the property of scale in the case of fractal calculus [44].

A function $g(S_{F^\epsilon}^{\alpha}) : F^\epsilon \to Re$ is called the fractal homogeneous of degree-$m\alpha$ or invariant under fractal rescaling, if we have

$$g(S_{F^\epsilon}^{\alpha}(\mu x)) = \mu^{m\alpha} g(S_{F^\epsilon}^{\alpha}(x)), \ \exists\, n, \ \forall\, \mu. \tag{4.24}$$

where we used

$$x \to \mu x \Rightarrow S_{F^\epsilon}^{\alpha}(\mu x) = \mu^{\alpha} S_{F^\epsilon}^{\alpha}(x). \tag{4.25}$$

For example, we consider the triadic Cantor set ($= F^\epsilon, \epsilon = 1/3$), $m = 1$, and $\mu = 1/3^n, n = 1, 2, \ldots$ Eq. (4.24) turns to

$$g(S_{F^\epsilon}^{\alpha}(\frac{1}{3^n}x)) = (\frac{1}{3^n})^{\alpha} g(S_{F^\epsilon}^{\alpha}(x)). \tag{4.26}$$

The local fractal derivative and its rescaling properties is given by

$$D_{F^\epsilon}^{\alpha} g(S_{F^\epsilon}^{\alpha}(\mu x)) = \mu^{m\alpha-\alpha} D_{F^\epsilon}^{\alpha} g(S_{F^\epsilon}^{\alpha}(x)). \tag{4.27}$$

4.7 Fractal Calculus and its Applications

In this section, we present mathematical models for the processes with the fractal time structure [40]. Space-time was considered as fractal sets which give new results and concepts [56–58].

4.7.1 Classical Mechanics on Fractal Sets

In this section, we review the Newton, the Lagrange, and the Hamilton mechanics on fractal time sets [45, 59]. **Newton's Laws on fractal time:** A point particle with mass m that moves in space with the fractal time. Newton's Second Law is given by:

$$f = m_K (D_{F^\epsilon}^{\alpha})^2 x(t), \quad t \in F^\epsilon, \tag{4.28}$$

where $x(t)$ is the position of particle, f is a force applied on particle and the physical dimension of $[m_K] = (Mass)(Time)^\alpha$, where $m_K = m$. Dynamics of the simple harmonic oscillator involving fractal time is suggested by:

$$-kx = m_{F^\epsilon} (D_{F^\epsilon}^{\alpha})^2 x(t), \ \ x(0) = A, \ \ D_{F^\epsilon}^{\alpha} x(t)|_0 = 0. \tag{4.29}$$

where k is the spring constant and $f = -kx$ is the force applied to the particle. The solution to Eq. (4.29) in view of conjugacy fractal calculus with the ordinary calculus is

$$x(t) = A\cos(\omega S_{F^\epsilon}^{\alpha}(t)), \tag{4.30}$$

where $\omega = \sqrt{k/m_{F^\epsilon}}$ is a constant. In view of upper bound of the staircase function, namely, $S_{F^\epsilon}^{\alpha}(t) < t^\alpha$, we have

$$x(t) \propto A\cos(\omega t^\alpha). \tag{4.31}$$

In Figure 4.2(b), we have plotted the position function of the particle Eq. (4.31) for different values of α.

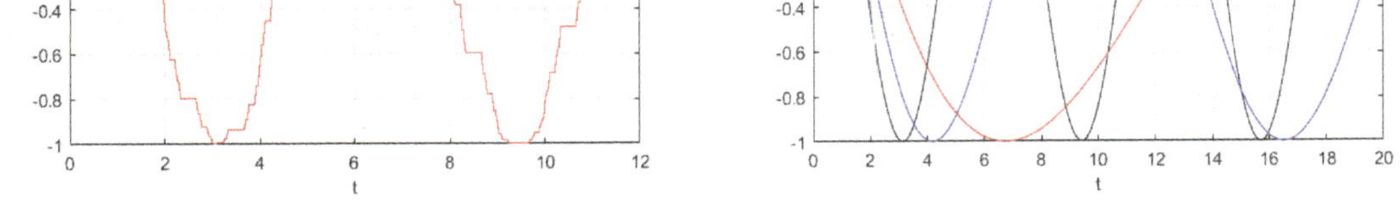

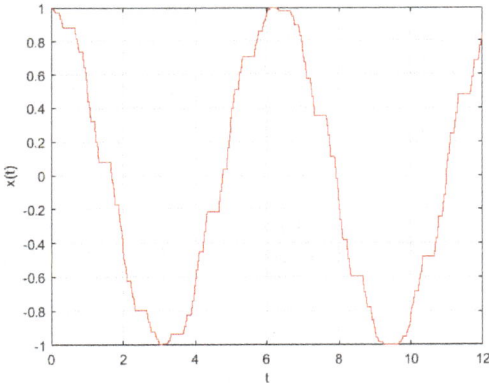

(a) Graph of Eq. (4.30) for the case $\epsilon = 1/3$.

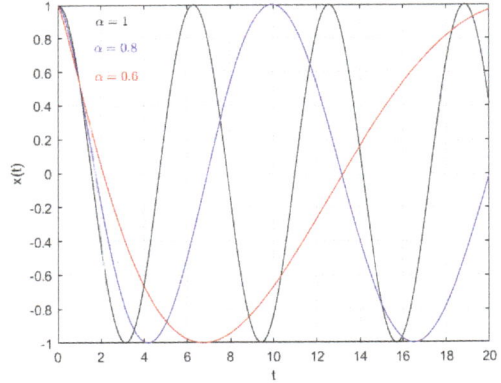

(b) Graph of Eq. (4.31) for different values of α.

Fig. 4.2 Graphs corresponding to simple harmonic oscillator.

Fractal Kepler's on fractal time: Let us consider the following fractal differential equation which is a mathematical model of the moving particle under force $f = -\kappa x(t)^{-2}$ as:

$$m_K (D_{F^\epsilon}^\alpha)^2 x(t) + \frac{\kappa}{x(t)^2} = 0. \tag{4.32}$$

where κ is a constant and $m_K = 1$. From scaling transformation of Eq. (4.32), we obtain

$$t \to Tt, \quad x \to \lambda x, \quad S_K^\alpha(Tt) \to T^\alpha S_K^\alpha(t)$$
$$D_{F^\epsilon}^\alpha \lambda x(Tt) \to \frac{\lambda}{T^\alpha} D_{F^\epsilon}^\alpha x(t). \tag{4.33}$$

According to the invariance of Eq. (4.32) under scale transform we get

$$T = \lambda^{3/2\alpha}, \tag{4.34}$$

where T is the orbital period and λ the semi-major axis of the particle moves on a elliptic orbit.

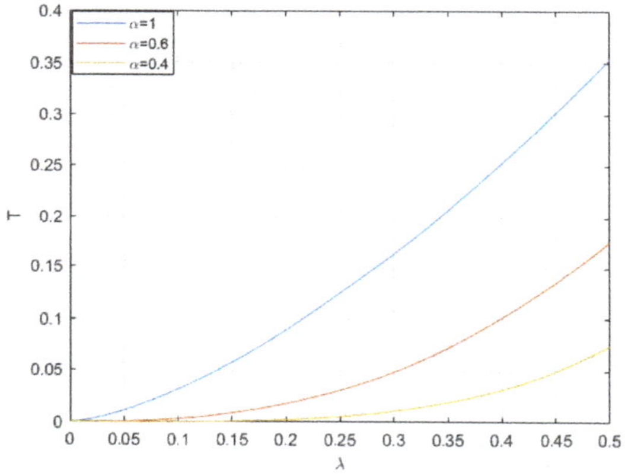

Fig. 4.3 Graph of the orbital period T versus the dimension of fractal time α.

In Figure 4.3, we have plotted Eq. (4.34) for different values of α. It shows that the orbital period T in the fractal time is smaller than standard case.

The Lagrangian and Hamiltonian on fractal time: Let $L(t, x(t), D_{F^\epsilon}^\alpha x) : F^\epsilon \times \mathrm{Re} \times \mathrm{Re} \to \mathrm{Re}$ be the Lagrangian of a particle on fractal time. The Euler-Lagrange equation corresponding to L has the following form [59]:

$$D_{F^\epsilon,x}^\alpha L - D_{F^\epsilon,t}^\alpha \frac{\partial L}{\partial D_{F^\epsilon,t}^\alpha x} = 0. \tag{4.35}$$

Using the fractal Legendre transform, it is easy to show that the fractal Hamilton equations are

$$D_{F^\epsilon,t}^\alpha x = \frac{\partial H}{\partial P_{F^\epsilon}}, \qquad D_{F^\epsilon,t}^\alpha P_{F^\epsilon} = D_{F^\epsilon,x}^\alpha H, \tag{4.36}$$

and

$$P_{F^\epsilon} = \frac{\partial L}{\partial D^{\alpha}_{F^\epsilon,t} x}. \tag{4.37}$$

where $D_{F^\epsilon,*}$ indicates the fractal derivatives respect to $*$ and P_{F^ϵ} might called fractal conjugate momentum.

4.7.2 Quantum Mechanics on Fractal Sets

In this section, the Schrödinger equation on fractal-time space is reviewed [47, 53].

Fractal Schrödinger equation on fractal time: The fractal Schrödinger equation involving fractal time for a moving particle with mass m is suggested as follows:

$$\frac{-\hbar^2}{2m}\frac{\partial^2 \psi(x,t)}{\partial x^2} + v(x)\psi(x,t) = i\hbar_{F^\epsilon} D^{\alpha}_{F^\epsilon,t}\psi(x,t), \quad x \in F^\epsilon, \tag{4.38}$$

where $[\hbar_{F^\epsilon}] = (Length^2)(Mass)/Time^{2-\alpha}$ the Plank constant ($\hbar_{F^\epsilon} = \hbar$), $v(x)$ the potential energy, and $\psi(x,t)$ is the wave function of the particle. The solution of Eq. (4.38) is

$$\begin{aligned} \psi(x,t) &\propto \phi(x) e^{\frac{-iES^{\alpha}_{F^\epsilon}(t)}{\hbar_{F^\epsilon}}} \\ &\propto \phi(x) e^{\frac{-iEt^{\alpha}}{\hbar_{F^\epsilon}}}, \end{aligned} \tag{4.39}$$

where E is the energy of particle.

Fractal Klein-Gordon equation on fractal time: The Klein-Gordon Equation on fractal-time is suggested by

$$-\frac{\zeta}{c^2}(D^{\alpha}_{F^\epsilon,t})^2\psi(x,t) + \frac{\partial^2\psi(x,t)}{\partial x^2} = \frac{m^2c^2}{\hbar^2}\psi(x,t), \tag{4.40}$$

where $[\zeta] = Time^{\alpha-1}$, ($\zeta = 1$) and m is the mass of the particle [60]. The solution of Eq. (4.40) is

$$\psi(x,t) \propto e^{ikx-\omega S^{\alpha}_{F^\epsilon}(t)}, \tag{4.41}$$

by using $S^{\alpha}_{F^\epsilon}(t) < t^{\alpha}$ we can write

$$\psi(x,t) \propto e^{ikx-\omega t^{\alpha}}, \tag{4.42}$$

where k is the wave number and $\omega = \sqrt{k^2c^2 + (m^2c^4/\hbar^2}$ is the angular frequency [60].

4.7.3 Diffraction Fringes from Fractal Sets

In this section, we review fractal gratings and its diffraction fringes in view of the Fraunhofer approximation model [49, 61–63]. In optics, the Fraunhofer diffraction equation is used to model the diffraction of waves when the diffraction pattern is viewed at a long distance from the diffracting object. The Fourier transform method is used to find the form of the diffraction for any periodic structure where the Fourier transform of the structure is defined. Here, we use the fractal Fourier transform for finding the diffraction pattern from fractal gratings. To do it we start by defining plane wave on a fractal set as follows:

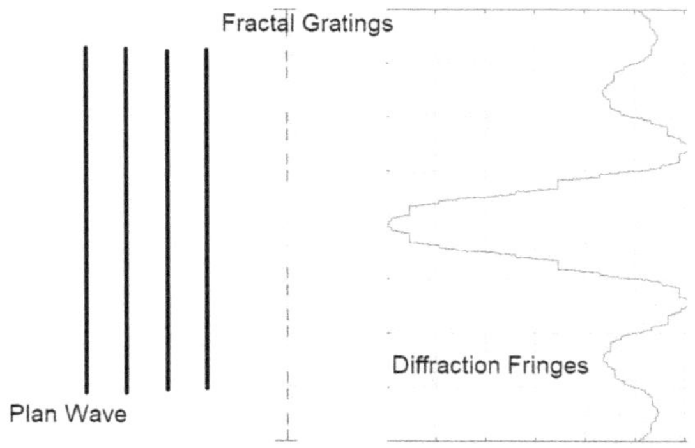

Fig. 4.4 The graph of diffraction pattern from fractal set.

$$f(t) = \begin{cases} e^{iS_{F\epsilon}^{\alpha}(\omega_0)S_{F\epsilon}^{\alpha}(t)}, & -\iota_0/2 < t < \iota_0/2; \\ 0, & \text{otherwise.} \end{cases} \tag{4.43}$$

where ι_0 is constant, ω_0 the frequency and t is the time.

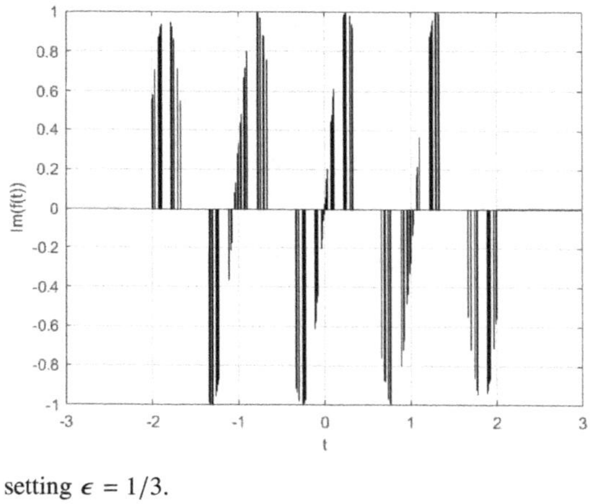

Fig. 4.5 Graph of real part of f setting $\epsilon = 1/3$.

In Figure 4.5, we have plotted real part wave function given by Eq. (4.43) for the case of $\epsilon = 1/3$ and $S_{F\epsilon}^{\alpha}(\omega_0) = 6$.
Using the fractal Fourier transform to Eq. (4.43), we obtain [49, 66]:

$$g(\omega) = \sqrt{\frac{2}{\pi}} \frac{\sin\left(S_{F\epsilon}^{\alpha}(\iota_0/2)(S_{F\epsilon}^{\alpha}(\omega) - S_{F\epsilon}^{\alpha}(\omega_0))\right)}{S_{F\epsilon}^{\alpha}(\omega) - S_{F\epsilon}^{\alpha}(\omega_0)} \tag{4.44}$$

In Figure 4.6, we have presented the diffraction fringes corresponding to the fractal set by choosing $\epsilon = 1/3$, and $\iota_0 = 2$.

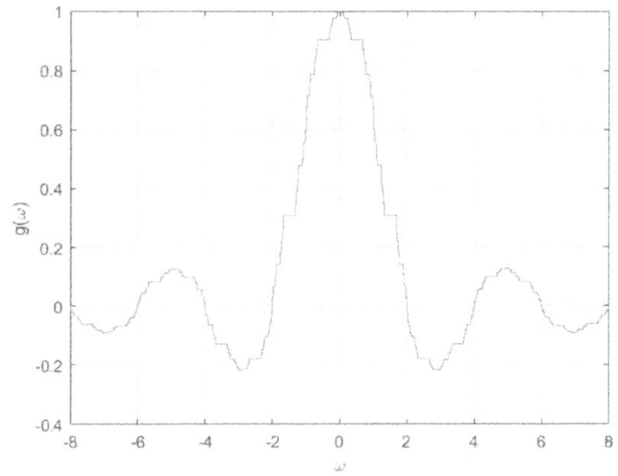

Fig. 4.6 Graph of g in the case of $\epsilon = 1/3$, $\iota_0 = 2$, and $\omega_0 = 0$.

Remark: We note that in all formulae of the chapter if we choose $\alpha = 1$, we obtain the standard results.

4.8 Concluding Remarks

By defining the F^α-continuity, F^α-derivative and F^α-integral for the thin Cantor set, the local fractal calculus has been discussed in this chapter. Further, graphs have been provided for the construction, Ψ-dimension, integral staircase function and, characteristic function of the thin Cantor set when $\epsilon = 1/2$ and various non-local fractal calculus formulae for the thin Cantor set has been defined. Fourier transform and local fractal Fourier transform for different functions have been tabulated. Besides, Laplace transform and local fractal Laplace transform for various functions have also been illustrated. Applications of fractal calculus have been discussed and their graphical representations are also demonstrated.

References

[1] Barnsley, M.F. 2014. Fractals Everywhere. Academic Press.
[2] Batty, M. and A.P. Longley. 1994. Fractal cities: A geometry of form and function. Academic Press.
[3] Bunde, A. and S. Havlin (eds.). 2013. Fractals in Science, Springer.
[4] Mandelbrot, B.B. 1977. The Fractal Geometry of Nature, Freeman and Company.
[5] Belair, J. and S. Dubuc (eds.). 2013. Fractal Geometry and Analysis (Vol. 346). Springer Science Business Media.
[6] Falconer, K. 1997. Techniques in Fractal Geometry. John Wiley and Sons.
[7] Edgar, G. 2007. Measure, Topology, and Fractal Geometry. Springer Science Business Media.
[8] Turner, M.J., J.M. Blackledge and P.R. Andrews. 1998. Fractal Geometry in Digital Imaging. Academic Press.
[9] Scholz, C.H. and B.B. Mandelbrot (eds.). 1989. Fractals in Geophysics (p. 313). Basel: Birkhauser Verlag.
[10] Banerjee, S., M.K. Hassan, S. Mukherjee and A. Gowrisankar. 2020. Fractal Patterns in Nonlinear Dynamics and Applications. CRC Press.

[11] Banerjee, S., D. Easwaramoorthy and A. Gowrisankar. 2021. Fractal Functions, Dimensions and Signal Analysis. Springer International Publishing.

[12] Fataf, N.A.A., A. Gowrisankar and S. Banerjee. 2020. In search of self-similar chaotic attractors based on fractal function with variable scaling approximately. Physica Scripta, 95(7): 075206.

[13] Prasad, P.K., A. Gowrisankar, A. Saha and S. Banerjee. 2020. Dynamical properties and fractal patterns of nonlinear waves in solar wind plasma. Physica Scripta, 95(6): 065603.

[14] Barlow, M.T. and E.A. Perkins. 1988. Brownian motion on the Sierpinski gasket. Probab. Theory Rel., 79(4): 543–623.

[15] Falconer, K. 1997. Techniques in Fractal Geometry, Wiley.

[16] Freiberg, U. and M. Zahle. 2002. Harmonic calculus on fractals-a measure geometric approach I. Potential Anal., 16: 265–277.

[17] Kigami, J. 2001. Analysis on Fractals. Cambridge University Press.

[18] Balankin, A.S. 2015. Effective degrees of freedom of a random walk on a fractal. Phys. Rev. E, 92: 062146.

[19] Hilfer, R. (ed.). 2000. Applications of Fractional Calculus in Physics. World Scientific.

[20] Uchaikin, V.V. 2013. Fractional Derivatives for Physicists and Engineers Vol. 1 Background and Theory. Vol. 2, Application Springer, Berlin.

[21] Tatom, F.B. 1995. The relationship between fractional calculus and fractals. Fractals, 3(1): 217–229.

[22] Priyanka, T.M.C. and A. Gowrisankar. 2021. Analysis on Weyl-Marchaud fractional derivative for types of fractal interpolation function with fractal dimension. Fractals, doi.org/10.1142/S0218348X21502157.

[23] Zubair, M., M.J. Mughal and Q.A. Naqvi. 2012. Electromagnetic Fields and Waves in Fractional Dimensional Space. Springer, New York.

[24] Herrmann, R. 2014. Fractional Calculus: An Introduction for Physicists, World Scientific.

[25] Tarasov, V.E. 2011. Fractional Dynamics: Applications of Fractional Calculus to Dynamics of Particles, Fields and Media. Springer Science Business Media.

[26] Das, S. 2011. Functional Fractional Calculus. Springer Science Business Media.

[27] Agarwal, R.P. and M. Bohner. 1999. Basic calculus on time scales and some of its applications. Results in Mathematics, 35(1-2): 3–22.

[28] Kunze, H., D. La Torre, F. Mendivil and E.R. Vrscay. 2019. Self-similarity of solutions to integral and differential equations with respect to a fractal measure. Fractals, 27(2): 950014.

[29] Czachor, M. 2019. Waves along fractal coastlines: From fractal arithmetic to wave equations. Acta Phys. Pol. B, 50: 813–831.

[30] Lopes, A.M. and J.A. Tenreiro Machado. 2018. Power law behaviour in complex systems. Entropy, 20(9): 1–3.

[31] Bies, A., C. Boydston, R. Taylor and M. Sereno. 2016. Relationship between fractal dimension and spectral scaling decay rate in computer-generated fractals. Symmetry, 8(7)66: 1–17.

[32] Samayoa, D., L. Alvarez-Romero, L.A. Ochoa-Ontiveros, L. Damián-Adame, E. Victoria-Tobon and G. Romero-Paredes. 2018. Fractal imbibition in Koch's curve-like capillary tubes. Rev. Mex. de Fís., 64(3): 291–295.

[33] Ri, S.I., V. Drakopoulos and S.M. Nam. 2021. Fractal interpolation using harmonic functions on the Koch curve. Fractal Fract., 5(28): 1–13.

[34] Parvate, A. and A.D. Gangal. 2009. Calculus on fractal subsets of real-line I: Formulation. Fractals, 17(01): 53–148.

[35] Parvate, A. and A.D. Gangal. 2011. Calculus on fractal subsets of real line II: Conjugacy with ordinary calculus. Fractals, 19(03): 271–290.

[36] Parvate, A., S. Satin and A.D. Gangal. 2011. Calculus on fractal curves in R^n. Fractals, 19(01): 15–27.

[37] Golmankhaneh, A.K. and A. Fernandez. 2008. Fractal calculus of functions on Cantor tartan spaces. Fractal Fract., 2(30): 1–13.

[38] Golmankhaneh, A.K., A. Fernandez, A.K. Golmankhaneh and D. Baleanu. 2018. Diffusion on middle-ξ Cantor sets. Entropy, 20(504): 1–13.

[39] Çetinkaya, F.A. and A.K. Golmankhaneh. 2021. General characteristics of a fractal Sturm-Liouville problem. Turk. J. Math., 45: 1835–1846.

[40] Golmankhaneh, A.K. and K. Welch. 2021. Equilibrium and non-equilibrium statistical mechanics with generalized fractal derivatives: A review. Mod. Phys. Lett. A, 36(14): 2140002.

[41] Golmankhaneh, A.K. and D. Baleanu. 2016. New derivatives on the fractal subset of Real-line. Entropy, 18(2): 1–13.

[42] Golmankhaneh, A.K. and D. Baleanu. 2016. Non-local integrals and derivatives on fractal sets with applications. Open Physics, 14(1): 542–548.

[43] Robert, D. and W. Urbina. 2014. On Cantor-like sets and Cantor-Lebesgue singular functions. arXiv preprint arXiv:1403.6554.

[44] Abed-Pour, N., A. Aghamohammadi, M. Khorrami and M.R.R. Tabar. 2003. Discrete scale invariance and its logarithmic extension. Nuclear Physics B, 655(3): 342–352.

[45] Golmankhaneh, A.K., V. Fazlollahi and D. Baleanu. 2013. Newtonian mechanics on fractals subset of real-line. Rom. Rep. Phys., 65(1): 84–93.

[46] Golmankhaneh, A.K., A.K. Golmankhaneh and D. Baleanu. 2013. Lagrangian and Hamiltonian mechanics on fractals subset of real-line. Int. J. Theor. Phys., 52(11): 4210–4217.

[47] Golmankhaneh, A.K., A.K. Golmankhaneh and D. Baleanu. 2015. About Schrödinger equation on fractals curves imbedding in R^3. Int. J. Theor. Phys., 54(4): 1275–1282.

[48] Golmankhaneh, A.K. and R.T. Sibatov. 2021. Fractal stochastic processes on thin Cantor-like sets. Mathematics, 9(6): 613.

[49] Golmankhaneh, A.K. and D. Baleanu. 2016. Diffraction from fractal grating Cantor sets. J. Mod. Opt., 63(14): 1364–1369.

[50] El-Nabulsi, R.A. and A.K. Golmankhaneh. 2021. On fractional and fractal Einstein's field equations. Mod. Phys. Lett. A, 36(05): 2150030.

[51] Golmankhaneh, A.K., A.K. Golmankhaneh and D. Baleanu. 2013. About Maxwell's equations on fractal subsets of R^3. Cent. Eur. J. Phys., 11(6): 863–867.

[52] Golmankhaneh, A.K. 2019. On the fractal Langevin equation. Fractal Fract., 3(1): 11, 1–9.

[53] Golmankhaneh, A.K. and C. Cattani. 2019. Fractal logistic equation. Fractal Fract., 3(3): 41, 1–13.

[54] Golmankhaneh, A.K. and C. Tunç. 2019. Sumudu transform in fractal calculus. Appl. Math. Comput., 350: 386–401.

[55] Golmankhaneh, A.K. and A.S. Balankin. 2018. Sub-and super-diffusion on Cantor sets: Beyond the paradox. Phys. Lett. A 382, 14: 960–967.

[56] Shlesinger, M.F. 1988. Fractal time in condensed mattar. Ann. Rev. Phys. Chern., 39: 269–290.

[57] Vrobel, S. 2011. Fractal time: Why a watched kettle never boils (Vol. 14). World Scientific.

[58] Welch, K. 2020. A Fractal Topology of Time: Deepening into Timelessness, Fox Finding Press. 2nd Edition.

[59] Golmankhaneh, A.K., A.K. Golmankhaneh and D. Baleanu. 2013. Lagrangian and Hamiltonian mechanics on fractals subset of real-line. Int. J. Theor. Phys., 52(11): 4210–4217.

[60] Srivastava, R.K. 2007. Statistical Mechanics. PHI Learning Pvt. Ltd.

[61] Lehman, M. 2001. Fractal diffraction gratings built through rectangular domains. Opt. Commun., 195(1-4): 11–26.

[62] Allain, C. and M. Cloitre. 1986. Optical diffraction on fractals. Phys. Rev. B, 33(5): 3566.
[63] Hou, B., G. Xu, W. Wen and G.K. Wong. 2004. Diffraction by an optical fractal grating. Appl. Phys. Lett., 85(25): 6125–6127.
[64] Monsoriu, J.A., W.D. Furlan, A. Pons, J.C. Barreiro and M.H. Giménez. 2011. Undergraduate experiment with fractal diffraction gratings. Eur. J. Phys., 32(3): 687.
[65] Teng, S., Y. Cui and Z. Li. 2015. Talbot effect of grating with fractal rough edges. J. Opt., 18(1): 015601.
[66] Golmankhaneh, A.K., K.K. Ali, R. Yilmazer and M.K.A. Kaabar. 2021. Local fractal fourier transform and applications. Comput. Methods Differ. Equ., doi:10.22034/cmde.2021.42554.1832.

Chapter 5

Perspective of Fractal Calculus on Types of Fractal Interpolation Functions

TMC Priyanka, A Agathiyan and A Gowrisankar*

5.1 Introduction

Recent years have seen a developing number of complex and nonlinear research problems which are challenging to be portrayed by traditional mathematical modeling, like Euclidean geometry and classical calculus. To remedy this irksome problem, diverse theoretical research notions, for instance, fractional calculus and fractal calculus, have been interconnected with fractal analysis since fractals play a vital role in the nonlinear system. Meanwhile, the concept of fractal interpolation functions has developed based on the iterated function systems very fast over the last two decades, which initiated a new research field in the context of interpolation and approximation. Numerous researchers working on the idea of the fractal interpolation function in the beginning stage has been shown in the article by Barnsley [17]. In contrast to classical methods, the fractal interpolation function produces complicated mathematical structures/naturally occurring functions with a simple recursive procedure, so the fractal approach provides more flexibility and versatility in approximation. Despite the approximation of continuous function done by polynomials, approximating irregular (non-smooth) function is significant as objects in the universe, in general, abound with the class of rough functions. Therefore, researchers give more attention on the problem of how to describe the non-smooth functions through fractal functions from different perspectives. Moreover, in order to extend the potential utility of FIFs and enlarge the flexibility in the interpolation theory, different types of fractal interpolation function are constructed. Especially, given a continuous function g defined on a compact interval on real line, suitable iterated function system is constructed to generate continuous functions g^{α} that simultaneously interpolate and approximate g (refer [2]).The notion of FIF extended to produce hidden variable fractal interpolation function (HFIF) which is non-self-affine and more irregular than a FIF [3–6]. The idea is to apply the FIF methodology to a generalized data set $\{(x_n, y_n, z_n) : n = 1, 2, \ldots, N\}$ in $\mathbb{R}^3$ such that the projection of the graph of the corresponding FIF onto $\mathbb{R}^2$ offers the required interpolation function for the data set $\{(x_n, y_n) : n = 1, 2, \ldots, N\}$.

There are continuous efforts which have been described in recent years to investigate fractal dimensions and the order of fractional calculus of non-smooth functions, for instance [16]–[31]. A general framework between fractional calculus and fractal functions has been narrated in [16]. In this text the linear connection between the order of the fractional calculus and the dimensions of the graph of the Weierstrass functions has investigated [17]. Further, the fractional integral of linear fractal interpolation function and its box dimension are explored by Ruan et al. in [18]. Though every one of these topics has expanded our understanding and brought out numerous excellent associations, an immediate and direct strategy including fractional order operators on fractals is just decently investigated. Despite the fact that the measure of the theoretical

Department of Mathematics, School of Advanced Sciences, Vellore Institute of Technology, Vellore 632 014, Tamil Nadu, India.
Emails: priyankamohan195@gmail.com; agathiyanbhc@gmail.com
* Corresponding author: gowrisankargri@gmail.com

approach is well crafted, Riemann integration-like procedures have their place. They are more straightforward, constructive and advantageous according to algorithmic perspective. Once in a while numerical calculations algorithms on Lebesgue integrals however are put together more regularly with respect to Riemann sums. Henceforth, recent research concentrates on measuring theoretical procedure. In particular, Parvate et al. have tuned to calculus to measure theoretical approach and systematically explained a series of calculus on fractal involving integrals and derivatives of appropriate orders in between 0 and 1 (refer [10–12]). Further, the geometrical explanation of fractal calculus has been presented by Ji-Huan He [19]. Chen et al. investigated the structural derivative approach, the implicit calculus equation modeling, and the fractal calculus operator in [20]. In this chapter, the fractal integral of the hidden variable fractal interpolation function and α-fractal function with the predefined initial condition is investigated.

The present work is organized as follows: Section 2 provides some theoretical background like the notions of FIF, CHFIF and α-fractal functions and revises some of their basic properties. The mathematical background of the fractal calculus elaborated in Section 3. In Section 4, fractal calculus of the hidden variable fractal interpolation function and α-fractal function is discussed. A conclusion is given in Section 5.

5.2 Fractal Interpolation Function

This section provides some theoretical background like the notions of FIF, CHFIF and α-fractal functions and revises some of their basic properties which are required for the subsequent sections. The interested reader is addressed to look at [3, 7–9, 17] for more details. Let us start this preliminary section by fixing the following notation which will be used throughout the article. For $N \in \mathbb{N}$, let $\mathbb{N}_N$ denote the subset $\{1, 2, \dots, N\}$ of $\mathbb{N}$, set of natural numbers.

Let a set of data points $\{(x_i, y_i) \in \mathbb{R}^2 : i \in \mathbb{N}_N\}$ such that $x_i < x_i + 1$ for all $i \in \mathbb{N}_{N-1}$ be given and are not necessarily equidistant. Let $I = [x_1, x_N]$, $I_i = [x_i, x_i + 1]$ for $i \in \mathbb{N}_{N-1}$ and $L_i : I \to I_i, i \in \mathbb{N}_{N-1}$ be $(N-1)$ contraction homeomorphisms such that

$$|L_i(x_1) - L_i(x_2)| \leq c_i|x_1 - x_2|,$$

where $c_i \in [0, 1)$ and satisfying the endpoint condition

$$L_i(x_1) = x_i, L_i(x_N) = x_{i+1}. \tag{5.1}$$

Let $r_i \in [0, 1), i \in \mathbb{N}_{N-1}$ and $X := I \times \mathbb{R}$. Let $N - 1$ continuous mappings $R_i : I \times \mathbb{R} \to \mathbb{R}$ satisfying the endpoint condition

$$R_i(x_1, y_1) = y_i, R_i(x_N, y_N) = y_{i+1}, \tag{5.2}$$

and R_i is contraction with respect to second variable. Thus, for all $y_1, y_2 \in \mathbb{R}$ and $x_1 \in I$,

$$|R_i(x_1, y_1) - R_i(x_1, y_2)| \leq r_i|y_1 - y_2|.$$

Define the iterated function system (IFS)

$$\{X; w_i : i \in \mathbb{N}_{N-1}\}, \tag{5.3}$$

where $w_i : X \to I_i \times \mathbb{R}, i \in \mathbb{N}_{N-1}$ given by

$$w_i(x, y) = (L_i(x), R_i(x, y)). \tag{5.4}$$

Associated with the IFS (5.3), there is a set valued map $W : \mathcal{K}(X) \to \mathcal{K}(X)$ defined by

$$W(B) = \bigcup_{i \in \mathbb{N}_{N-1}} w_i(B),$$

for any $B \in \mathcal{K}(X)$, set of all non-empty compact subsets of X. The Hausdorff metric completes $\mathcal{K}(X)$, since X is complete. Moreover, W is contraction on $\mathcal{K}(X)$ hence by the theory of IFS, W has a unique compact set G such that $G = W(G)$ and G is the graph of a continuous function $g : I \to \mathbb{R}$ which yields $g(x_i) = y_i$ for $i \in \mathbb{N}_N$. The function g whose graph is the attractor of an IFS (5.3) is called a Fractal Interpolation Function (FIF) or fractal function associated with the given data set $\{(x_i, y_i) \in \mathbb{R}^2 : i \in \mathbb{N}_N\}$.

The construction process of an iterated function system in $\mathbb{R}^2$ such that its attractor is graph of the interpolation function of given data, is explained by defining the Read-Bajraktarević (RB) operator $T : \mathbb{G} \to \mathbb{G}$ by

$$(Tf)(x) = R_i(L_i^{-1}(x), f \circ L_i^{-1}(x)),\ x \in I_i,\ i \in \mathbb{N}_{N-1}, \tag{5.5}$$

where $\mathbb{G} = \{f : I \to \mathbb{R} | f \text{ is continuous on } I, f(x_1) = y_1, f(x_N) = y_N\}$ is complete metric space endowed with the uniform metric $\delta(f, f') = \max\{|f(x) - f'(x)| : x \in I\}$. Then T is a contraction mapping on $(\mathbb{G}, \delta)$ with a contraction ratio $r = \max\{r_i : i \in \mathbb{N}_{N-1}\} < 1$. According to the theory of IFS, T has a unique fixed point, say g, which is the FIF corresponding to the IFS (5.3). Therefore, g satisfies the functional equation

$$g(x) = R_i(L_i^{-1}(x), g \circ L_i^{-1}(x)),\ x \in I_i,\ i \in \mathbb{N}_{N-1}. \tag{5.6}$$

or

$$g(L_i(x)) = R_i(x, g(x)),\ x \in I_i,\ i \in \mathbb{N}_{N-1}.$$

The most extensively studied FIF is generated from the IFS of the form

$$L_i(x) = a_i x + b_i,\ R_i(x, y) = \alpha_i y + q_i(x),\ i \in \mathbb{N}_{N-1}, \tag{5.7}$$

where $\{\alpha_i : \alpha_i \in (-1, 1), i \in \mathbb{N}_{N-1}\}$ is a family of parameters named as vertical scaling factors of the transformation w_i, and $\alpha = (\alpha_1, \alpha_2, \ldots, \alpha_{N-1})$ is the scale vector corresponding to the IFS (5.7) and $q_i : I \to \mathbb{R}$ is a continuous function satisfying

$$q_i(x_1) = y_i - \alpha_i y_1,\ q_i(x_N) = y_{i+1} - \alpha_i y_N, \text{ for each } i \in \mathbb{N}_{N-1}.$$

Example 5.1 Let a dataset $\{(0, 0), (2/4, 2/5), (3/4, 1/5), (1, 0)\}$ be given with vertical scaling factors $\alpha_1 = 1/2$, $\alpha_2 = -1/2$ and $\alpha_3 = 1/2$. The fractal interpolation function g is determined by the IFS consisting the following functions:

$$L_1(x) = \frac{1}{2}x, R_1(x, y) = \frac{1}{2}y + \frac{2}{5}x,$$
$$L_2(x) = \frac{1}{4}x + \frac{1}{2}, R_2(x, y) = \frac{-1}{2}y - \frac{1}{5}x + \frac{2}{5},$$
$$L_3(x) = \frac{1}{4}x + \frac{3}{4}, R_3(x, y) = \frac{1}{2}y - \frac{1}{5}x + \frac{1}{5}.$$

The graphical representation of g is depicted in Figure 5.1.

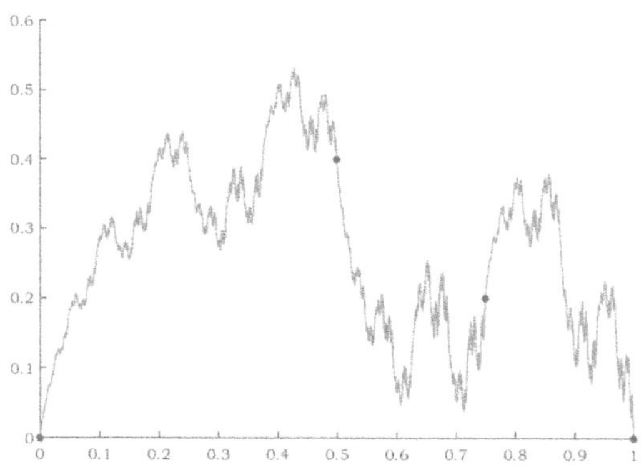

Fig. 5.1 Fractal interpolation function of the data points $\{(0,0), (2/4, 2/5), (3/4, 1/5), (1, 0)\}$ with the vertical scaling factors $\alpha_1 = 1/2$, $\alpha_2 = -1/2$, $\alpha_3 = 1/2$.

5.3 Hidden Variable Fractal Interpolation Function

This section succinctly reviews the hidden variable FIFs (HFIFs) which are constructed by projecting vector valued FIF corresponding to a generalized interpolation data which approximate non self- affine patterns.

Consider a generalized data set $\widehat{D} = \{(x_i, y_i, z_i) \in I \times \mathbb{R}^2 : i \in \mathbb{N}_N\}$ where $\{z_i : i \in \mathbb{N}_N\}$ are real parameters. The idea is to construct a fractal interpolation function $g_1 : I \to \mathbb{R}$ such that $g_1(x_i) = y_i$ for all $i \in \mathbb{N}_N$, project its graph into $I \times \mathbb{R}$ such a way that the projection is the graph of a function that interpolates $\{(x_i, y_i)\}$. Let $\mathbb{R}^2$ be endowed with the Manhattan metric $d_M((x_1, y_1), (x_2, y_2)) = |x_1 - x_2| + |y_1 - y_2|$, which is induced by the l^1-norm. For $i \in \mathbb{N}_{N-1}$, let the contraction homeomorphisms $L_i : I \to I_i \subset I$ be as given in (5.1), and the functions $R_i : I \times \mathbb{R}^2 \to \mathbb{R}^2$ be expressed as

$$R_i(x, \mathbf{y}) = R_i(x, y, z) = (R_i^1(x, y, z), R_i^2(x, z))^t := A_i(y, z)^t + (p_i(x), q_i(x))^t, \tag{5.8}$$

where t denotes the transpose, A_n are upper-triangular matrices $\begin{bmatrix} \alpha_i & \beta_i \\ 0 & \gamma_i \end{bmatrix}$, and p_i, q_i are suitable real valued continuous functions so that the following conditions are satisfied for all $i \in \mathbb{N}_{N-1}$:
(i) $d_M(R_i(x, y, z), R_i(x^*, y, z)) \leq c_1 |x - x^*|$ for some constant $c_1 > 0$,
(ii) $d_M(R_i(x, y, z), R_i(x, y^*, z^*)) \leq s\, d_M((y, z)(y^*, z^*))$ for $0 \leq s < 1$,
(iii) $R_i(x_1, y_1, z_1) = (y_i, z_i)$ and $R_i(x_N, y_N, z_N) = (y_{i+1}, z_{i+1})$.
The variables α_i, β_i, and γ_i are chosen such that $||A_i||_1 < 1$ for all $i \in \mathbb{N}_{N-1}$. Define $w_i : I \times \mathbb{R}^2 \to I \times \mathbb{R}^2$ by

$$w_i(x, y, z) = (L_i(x), R_i(x, y, z)). \tag{5.9}$$

It follows from the conditions on L_i and F_i that w_i are contraction maps with respect to the metric d_M^* defined on $I \times \mathbb{R}^2$ by $d_M^*((x, y, z), (x^*, y^*, z^*)) = |x - x^*| + \theta d_M((y, z), (y^*, z^*))$, where $\theta = \frac{1-a}{2c_1}$, and $a = \max\{\frac{h_i}{x_N - x_1}; i \in \mathbb{N}_{N-1}\}$, where $h_i = x_{i+1} - x_i$. Consequently, the generalized IFS $\{I \times \mathbb{R}^2; w_i : i \in \mathbb{N}_{N-1}\}$ admits an attractor $A \subseteq I \times \mathbb{R}^2$. It follows from the generalized IFS theory that the aforementioned attractor A is the graph of the continuous vector-valued function $\mathbf{g} : I \to \mathbb{R}^2$ such that $\mathbf{g}(x_i) = (y_i, z_i)$ for all $i \in \mathbb{N}_{N-1}$. Letting $\mathbf{g} = (g_1, g_2)$ it follows that $g_1 : I \to \mathbb{R}$ is a continuous function interpolating (x_i, y_i) and is called (coalescence) hidden variable fractal interpolation function (CHFIF) (see, for instance, [5]). Similarly, the projection $\{(x, g_2(x)) : x \in I\}$ of the attractor A is self-affine and provides a fractal function g_2 that interpolates the data $\{(x_i, z_i) : i \in \mathbb{N}_N\}$.

Let $\mathbb{G}^*$ be the set of continuous functions $\mathbf{h} : I \to \mathbb{R}^2$ such that $\mathbf{h}(x_1) = (y_1, z_1), \mathbf{h}(x_N) = (y_N, z_N)$ equipped with the metric $d(\mathbf{h}, \mathbf{h}^*) = \max\{d_M(\mathbf{h}(x), \mathbf{h}^*(x)) : x \in I\}$. To obtain a functional equation for g, we recall that g is the fixed point of the operator $T^* : \mathbb{G}^* \to \mathbb{G}^*$ defined by

$$(T^*h)(x) = R_i(L_i^{-1}(x), h(L_i^{-1}(x))), \text{for } x \in I_i, i \in \mathbb{N}_{N-1}.$$

Whence, the vector-valued function $\mathbf{g}$ satisfies the functional equation

$$\mathbf{g}(L_i(x)) = A_i\mathbf{g}(x) + (p_i(x), q_i(x))^t, x \in I. \tag{5.10}$$

The image $T^*\mathbf{g}$ of the vector-valued function $\mathbf{g} = (g_1, g_2)$ can be written componentwise as (T_1g_1, T_2g_2), where T_1 and T_2 are the componentwise Read-Bajraktarević operator. Then satisfy

$$\begin{aligned} g_1(L_i(x)) = T_1g_1(L_i(x)) = R_i^1(x, g_1(x), g_2(x)) = \alpha_i g_1(x) + \beta_i g_2(x) + p_i(x), \\ g_2(L_i(x)) = T_2g_2(L_i(x)) = R_i^2(x, g_2(x)) = \gamma_i g_2(x) + q_i(x), x \in I. \end{aligned} \tag{5.11}$$

Example 5.2 Let a dataset $\{(0, 0, 0), (2/4, 1, 2), (3/4, -1, 8), (1, 2, 5)\}$ be given. The hidden variable fractal interpolation function g is determined by the IFS consisting the following functions:

$$\begin{aligned} L_1(x) &= \frac{1}{2}x, R_1^1(x, y, z) = \frac{1}{2}y + \frac{1}{3}z - \frac{2}{3}x - 1, R_1^2(x, z) = \frac{1}{2}z - 2x + \frac{3}{2}, \\ L_2(x) &= \frac{1}{4}x + \frac{1}{2}, R_2^1(x, y, z) = \frac{1}{2}y + \frac{1}{3}z - \frac{11}{3}x, R_2^2(x, z) = \frac{1}{2}z + 5x + \frac{1}{2}, \\ L_3(x) &= \frac{1}{4}x + \frac{3}{4}, R_3^1(x, y, z) = \frac{1}{2}y + \frac{1}{3}z + \frac{4}{3}x - 2, R_3^2(x, z) = \frac{1}{2}z - 4x + \frac{13}{2}. \end{aligned}$$

The graphical representation of non-self-affine fractal function g_1 and self-affine fractal interpolation function g_2 are depicted in Figure 5.2(a), Figure 5.2(b) respectively.

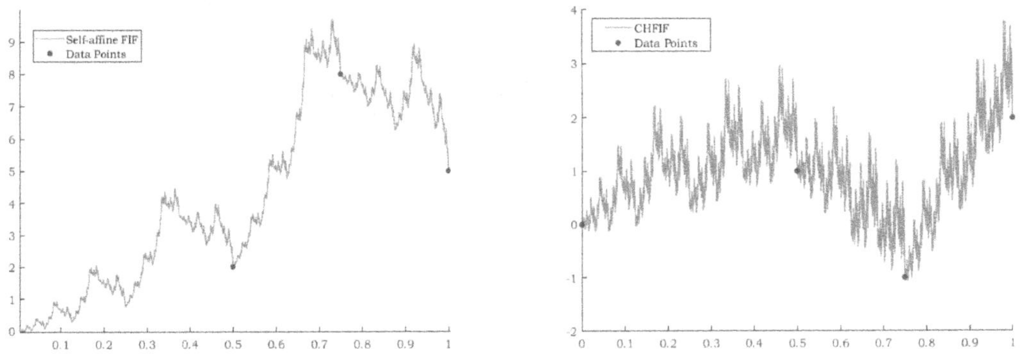

(a) Non-self-affine

(b) Self-affine

Fig. 5.2 Hidden variable fractal interpolation function.

5.4 α-Fractal Functions

Let $f \in C(I)$ be a continuous function and consider the case:

$$q_i(x) = f \circ L_i(x) - \alpha_i b(x). \tag{5.12}$$

Here $b : I \to \mathbb{R}$ is a continuous map that fulfills the conditions $b(x_1) = f(x_1)$, and $b(x_N) = f(x_N)$ and $b \neq f$. This case is proposed by Barnsley [1] and Navascués [2] as generalization of any continuous function. We define the α-fractal function corresponding to f in the following:

Definition 5.1 The continuous function $f^\alpha : I \to \mathbb{R}$ whose graph is the attractor of the IFS defined by (5.7),(5.12) is referred to as α-fractal function associated with f , with respect to b and the partition D.

$$L_i(x) = a_i x + b_i, \; R_i(x, y) = \alpha_i y + q_i(x), \; i \in \mathbb{N}_{N-1}, \tag{5.13}$$

where $q_i(x)$ is given in 5.12. According to (5.6), f^α satisfies the functional equation:

$$f^\alpha(x) = R_i(x, f^\alpha(L_i^{-1}(x))) = f(x) + \alpha_i[(f^\alpha - b) \circ L_i^{-1}(x)], \; \forall x \in I_i, i \in \mathbb{N}_{N-1}. \tag{5.14}$$

Note that for $\alpha = 0$, $f^\alpha = f$. Thus aforementioned equation may be treated as an entire family of functions f^α with f as its germ. By this method one can define fractal analogues of any continuous function.

5.4.1 Fractal Calculus

In this section we briefly recall the requisite general material of fractal calculus. For a detailed exposition the reader may refer to [10]–[15].

A fractal curve is defined as an image of continuous functions f from $\mathbb{R}$ to $\mathbb{R}^2$ which are fractals. A fractal curve $F \subset \mathbb{R}^n$ is said to be parameterizable if there exists a function $w : \mathbb{R} \to F$ which is continuous, one-to-one and onto. A subdivision $\mathbf{P}_{[a,b]}$ (or simply $\mathbf{P}$) of the interval $[a, b]$ is a finite set of points $\{a = x_0, x_1, \ldots, x_n = b\}, x_i < x_{i+1}$. Any interval of the form $[x_i, x_{i+1}]$ is called a component of the subdivision P. Let F be a fractal curve and $\mathbf{P}_{[a,b]}$ be a subdivision. The mass function is defined as

$$\gamma^\alpha(F, a, b) = \lim_{\delta \to 0} \inf_{\{P_{[a,b]}:|P| \leq \delta\}} \sum_{i=1}^{n-1} \frac{|w(t_i) - w(t_{i+1})|^\alpha}{\Gamma(\alpha + 1)}, \tag{5.15}$$

where $|.|$ denotes the Euclidean metric on R^3 and

$$|P| = \max_{0 \leq i \leq n-1} (t_{i+1} - t_i)$$

for a subdivision P. The staircase functions for the fractal curve is defined by

$$S_F^\alpha(t) = \begin{cases} \gamma^\alpha(F, p_0, t), & t \geq p_0, \\ -\gamma^\alpha(F, t, p_0), & t < p_0, \end{cases} \tag{5.16}$$

where $p_0 \in [a, b]$ is an arbitrary and fixed point.

Definition 5.2 The γ-dimension of the fractal curve F is defined as

$$\dim_\gamma(F) = \inf\{\alpha : \gamma^\alpha(F,a,b) = 0\} = \sup\{\alpha : \gamma^\alpha(F,a,b) = \infty\}. \tag{5.17}$$

Definition 5.3 The F^α-derivative of a function f at $\theta \in F$ is defined as

$$D_F^\alpha f(\theta) = F - \lim_{\theta' \to \theta} \frac{f(\theta') - f(\theta)}{J(\theta') - J(\theta)}, \tag{5.18}$$

where $J(\theta) = S_F^\alpha(w^{-1}(\theta)), \theta \in F$ and if the limit exists.

A segment $C(t_1, t_2)$ of the fractal curve is defined as

$$C(t_1, t_2) = \{w(t') : t' \in [t_1, t_2]\}. \tag{5.19}$$

Define

$$M[f, C(t_1, t_2)] = \sup_{\theta \in C(t_1, t_2)} f(\theta)$$

and

$$m[f, C(t_1, t_2)] = \inf_{\theta \in C(t_1, t_2)} f(\theta)$$

Definition 5.4 The upper and the lower f^α-sum for the function f over the subdivision P are defined as

$$U^\alpha[f, F, P] = \sum_{i=0}^{n-1} M[f, C(t_i, t_{i+1})][S_F^\alpha(t_{i+1}) - S_F^\alpha(t_i)] \tag{5.20}$$

$$L^\alpha[f, F, P] = \sum_{i=0}^{n-1} m[f, C(t_i, t_{i+1})][S_F^\alpha(t_{i+1}) - S_F^\alpha(t_i)]. \tag{5.21}$$

Define

$$\underline{\int_{C(a,b)}} f(\theta) d_F^\alpha \theta = \sup_{P[a,b]} L^\alpha[f, F, P]$$

and

$$\overline{\int_{C(a,b)}} f(\theta) d_F^\alpha \theta = \inf_{P[a,b]} U^\alpha[f, F, P].$$

Definition 5.5 The F^α-integral of the function f is defined as

$$\int_{C(a,b)} f(\theta) d_F^\alpha \theta = \underline{\int_{C(a,b)}} f(\theta) d_F^\alpha \theta = \overline{\int_{C(a,b)}} f(\theta) d_F^\alpha \theta. \tag{5.22}$$

5.5 Fractal Calculus of Fractal Functions

Let g be the hidden variable fractal interpolation function generated by the IFS provided in Eq. (5.9). The vector-valued function $\mathbf{g}$ satisfies the functional equation given in Eq. (5.10). The image $T^*\mathbf{g}$ of the vector-valued function $\mathbf{g} = (g_1, g_2)$ can be written componentwise as $(T_1 g_1, T_2 g_2)$, where T_1 and T_2 are satisfying the Eq. (5.11). For given $\tilde{y}_1, \tilde{z}_1$ the fractal integral of order of β of $\mathbf{g} = (g_1, g_2)$ is defined as follow

$$\tilde{g}_1(x) = \tilde{y}_1 + \int_{x_1}^{x} S_F^{\beta}(g_1(t))d_f^{\beta}t. \tag{5.23}$$

$$\tilde{g}_2(x) = \tilde{z}_1 + \int_{x_1}^{x} S_F^{\beta}(g_2(t))d_f^{\beta}t. \tag{5.24}$$

Theorem 5.1 *Let* $\mathbf{g}$ *be the hidden variable fractal interpolation function generated by the IFS provided in Eq. (5.9). For given* $\tilde{y}_1, \tilde{z}_1$, *fractal integral of* $\mathbf{g}$ *is defined in Eq. (5.23), Eq. (5.24), denoted as* $\tilde{\mathbf{g}}$. *Then* $\tilde{\mathbf{g}} = (\tilde{g}_1, \tilde{g}_2)$ *is also hidden variable fractal interpolation function associated with* $\{(L_i(x), \tilde{R}_i(x, y, z)) : i \in \mathbb{N}_{N-1}\}$, *where, for* $i \in \mathbb{N}_{N-1}$,

$$\begin{aligned}
\tilde{R}_i &= \tilde{A}_i(y, z)^t + (\tilde{p}_i(x), \tilde{q}_i(x))^t \\
\tilde{A}_i &= a_i \begin{bmatrix} \alpha_i & \beta_i \\ 0 & \gamma_i \end{bmatrix}, \\
\tilde{p}_i(x) &= \tilde{y}_i - a_i\alpha_i\tilde{y}_1 - a_i\beta_i\tilde{z}_1 + a_i \int_{x_1}^{x} S_F^{\beta}(p_i(t))d_F^{\beta}t, \\
\tilde{y}_{i+1} &= \tilde{y}_1 + \sum_{k=1}^{i} a_k \left[\alpha_k(\tilde{y}_N - \tilde{y}_1) + \beta_k(\tilde{z}_N - \tilde{z}_1) + \int_{x_1}^{x_N} S_F^{\beta}(p_i(t))d_F^{\beta}t\right], \\
\tilde{y}_N &= \tilde{y}_1 + \frac{\sum_{k=1}^{N-1} a_k \left[\beta_k(\tilde{z}_N - \tilde{z}_1) + \int_{x_1}^{x_N} S_F^{\beta}(p_k(t))d_F^{\beta}t\right]}{1 - \sum_{k=1}^{N-1} a_k\alpha_k}, \\
\tilde{q}_i(x) &= \tilde{z}_i - a_i\gamma_i\tilde{z}_1 + a_i \int_{x_1}^{x} S_F^{\beta}(q_i(t))d_F^{\beta}t, \\
\tilde{z}_{i+1} &= \tilde{z}_1 + \sum_{k=1}^{i} a_k \left[\gamma_k(\tilde{z}_N - \tilde{z}_1) + \int_{x_1}^{x_N} S_F^{\beta}(q_k(t))d_F^{\beta}t\right], \\
\tilde{z}_N &= \tilde{z}_1 + \frac{\sum_{k=1}^{N-1} a_k \int_{x_1}^{x_N} S_F^{\beta}(q_k(t))d_F^{\beta}t}{1 - \sum_{k=1}^{N-1} a_k\gamma_k}.
\end{aligned}$$

Proof: Applying the definition of fractal integral on the g_1 function gives

$$\begin{aligned}
\tilde{g}_1(L_i(x)) &= \tilde{y}_1 + \int_{x_1}^{L_i(x)} S_F^{\beta}(g_1(t))d_F^{\beta}t \\
&= \tilde{y}_1 + \int_{x_1}^{x_i} S_F^{\beta}(g_1(t))d_F^{\beta}t + \int_{x_i}^{L_i(x)} S_F^{\beta}(g_1(t))d_F^{\beta}t \\
&= \tilde{y}_i + a_i \int_{x_1}^{x} S_F^{\beta}(g_1(L_i(t)))d_F^{\beta}t.
\end{aligned}$$

Since, the component of the hidden variable fractal function g_1 satisfies the functional equation given in Eq. (5.11)

$$\begin{aligned}\tilde{g}_1(L_i(x)) &= \tilde{y}_i + a_i \int_{x_1}^{x} S_F^{\beta}(\alpha_i g_1(t) + \beta_i g_2(t) + p_i(t)) d_F^{\beta} t \\ &= \tilde{y}_i + a_i\alpha_i \int_{x_1}^{x} S_F^{\beta}(g_1(t)) d_F^{\beta} t + a_i\beta_i \int_{x_1}^{x} S_F^{\beta}(g_2(t)) d_F^{\beta} t + a_i \int_{x_1}^{x} S_F^{\beta}(p_i(t)) d_F^{\beta} t.\end{aligned}$$

Equation (5.23) and Eq. (5.24) provide

$$\begin{aligned}\tilde{g}_1(L_i(x)) &= \tilde{y}_i + a_i\alpha_i(\tilde{g}_1(x) - \tilde{y}_1) + a_i\beta_i(\tilde{g}_2(x) - \tilde{z}_1) + a_i \int_{x_1}^{x} S_F^{\beta}(p_i(t)) d_F^{\beta} t. \\ &= a_i\alpha_i \tilde{g}_1(x) + a_i\beta_i \tilde{g}_2(x) + \tilde{p}_i(x),\end{aligned}$$

where $\tilde{p}_i(x) = \tilde{y}_i - a_i\alpha_i\tilde{y}_1 - a_i\beta_i\tilde{z}_1 + a_i \int_{x_1}^{x} S_F^{\beta}(p_i(t)) d_F^{\beta} t$. Similarly,

$$\begin{aligned}\tilde{g}_2(L_i(x)) &= \tilde{z}_1 + \int_{x_1}^{L_i(x)} S_F^{\beta}(g_2(t)) d_F^{\beta} t \\ &= \tilde{z}_1 + \int_{x_1}^{x_i} S_F^{\beta}(g_2(t)) d_F^{\beta} t + \int_{x_i}^{L_i(x)} S_F^{\beta}(g_2(t)) d_F^{\beta} t \\ &= \tilde{z}_i + a_i \int_{x_1}^{x} S_F^{\beta}(g_2(L_i(t))) d_F^{\beta} t.\end{aligned}$$

Since, g_2 is the component of hidden variable fractal function and satisfies the functional equation given in Eq. (5.11)

$$\begin{aligned}\tilde{g}_2(L_i(x)) &= \tilde{z}_i + a_i \int_{x_1}^{x} S_F^{\beta}(\gamma_i g_2(t) + q_i(t)) d_F^{\beta} t \\ &= \tilde{z}_i + a_i\gamma_i \int_{x_1}^{x} S_F^{\beta}(g_2(t)) d_F^{\beta} t + a_i \int_{x_1}^{x} S_F^{\beta}(q_i(t)) d_F^{\beta} t.\end{aligned}$$

Equation (5.24) gives

$$\begin{aligned}\tilde{g}_2(L_i(x)) &= \tilde{z}_i + a_i\gamma_i(\tilde{g}_2(x) - \tilde{z}_1) + a_i \int_{x_1}^{x} S_F^{\beta}(q_i(t)) d_F^{\beta} t. \\ &= a_i\gamma_i \tilde{g}_2(x) + \tilde{q}_i(x),\end{aligned}$$

where $\tilde{q}_i(x) = \tilde{z}_i - a_i\gamma_i\tilde{z}_1 + a_i \int_{x_1}^{x} S_F^{\beta}(q_i(t)) d_F^{\beta} t$. Hence, $\tilde{\mathbf{g}} = (\tilde{g}_1, \tilde{g}_2)$ is hidden variable fractal interpolation function generated by the IFS $\{(L_i(x), \tilde{R}_i(x, y, z)) : i \in \mathbb{N}_{N-1}\}$. Take $x = x_N$, $L_i(x_N) = x_{i+1}$, then

$$\tilde{y}_{i+1} = \tilde{y}_i + a_i\alpha_i(\tilde{y}_N - \tilde{y}_1) + a_i\beta_i(\tilde{z}_N - \tilde{z}_1) + a_i \int_{x_1}^{x_N} S_F^\beta(p_i(t))d_F^\beta t$$

$$\tilde{y}_{i+1} - \tilde{y}_i = a_i \left[\alpha_i(\tilde{y}_N - \tilde{y}_1) + \beta_i(\tilde{z}_N - \tilde{z}_1) + \int_{x_1}^{x_N} S_F^\beta(p_i(t))d_F^\beta t\right]$$

$$\tilde{z}_{i+1} = \tilde{z}_i + a_i\gamma_i(\tilde{z}_N - \tilde{z}_1) + a_i \int_{x_1}^{x_N} S_F^\beta(q_i(t))d_F^\beta t$$

$$\tilde{z}_{i+1} - \tilde{z}_i = a_i\gamma_i(\tilde{z}_N - \tilde{z}_1) + a_i \int_{x_1}^{x_N} S_F^\beta(q_i(t))d_F^\beta t.$$

Clearly, $\tilde{y}_{i+1} = \tilde{y}_1 + \sum_{k=1}^{i}(\tilde{y}_{k+1} - \tilde{y}_k)$, this equation yields the new data points

$$\tilde{y}_{i+1} = \tilde{y}_1 + \sum_{k=1}^{i} a_k \left[\alpha_k(\tilde{y}_N - \tilde{y}_1) + \beta_k(\tilde{z}_N - \tilde{z}_1) + \int_{x_1}^{x_N} S_F^\beta(p_k(t))d_F^\beta t\right]. \tag{5.25}$$

$$\tilde{z}_{i+1} = \tilde{z}_1 + \sum_{k=1}^{i} a_k \left[\gamma_k(\tilde{z}_N - \tilde{z}_1) + \int_{x_1}^{x_N} S_F^\beta(p_k(t))d_F^\beta t\right]. \tag{5.26}$$

Taking $i = N - 1$ in Eq. (5.25) and Eq. (5.26) give the endpoints $\tilde{y}_N, \tilde{z}_N$ of the new data set

$$\tilde{y}_N = \tilde{y}_1 + \sum_{k=1}^{N-1} a_k \left[\alpha_k(\tilde{y}_N - \tilde{y}_1) + \beta_k(\tilde{z}_N - \tilde{z}_1) + \int_{x_1}^{x_N} S_F^\beta(p_k(t))d_F^\beta t\right]$$

$$(\tilde{y}_N - \tilde{y}_1)\left[1 - \sum_{k=1}^{N-1} a_k\alpha_k\right] = \sum_{k=1}^{N-1} a_k \left[\beta_k(\tilde{z}_N - \tilde{z}_1) + \int_{x_1}^{x_N} S_F^\beta(p_k(t))d_F^\beta t\right].$$

A similar argument gives the endpoint $\tilde{z}_N$. Thus, $\tilde{\mathbf{g}} = (\tilde{g}_1, \tilde{g}_2)$ is hidden variable fractal interpolation function which interpolates the new set of data $\{(x_i, \tilde{y}_i, \tilde{z}_i) : i \in \mathbb{N}_N\}$, where, for each $i \in \mathbb{N}_{N-1}$

$$\tilde{y}_{i+1} = \tilde{y}_1 + \sum_{k=1}^{i} a_k \left[\alpha_k(\tilde{y}_N - \tilde{y}_1) + \beta_k(\tilde{z}_N - \tilde{z}_1) + \int_{x_1}^{x_N} S_F^\beta(p_k(t))d_F^\beta t\right],$$

$$\tilde{z}_{i+1} = \tilde{z}_1 + \sum_{k=1}^{i} a_k \left[\gamma_k(\tilde{z}_N - \tilde{z}_1) + \int_{x_1}^{x_N} S_F^\beta(p_k(t))d_F^\beta t\right],$$

$$\tilde{y}_N = \tilde{y}_1 + \frac{\sum_{k=1}^{N-1} a_k \left[\beta_k(\tilde{z}_N - \tilde{z}_1) + \int_{x_1}^{x_N} S_F^\beta(p_k(t))d_F^\beta t\right]}{1 - \sum_{k=1}^{N-1} a_k\alpha_k},$$

$$\tilde{z}_N = \tilde{z}_1 + \frac{\sum_{k=1}^{N-1} a_k \left[\int_{x_1}^{x_N} S_F^\beta(q_k(t))d_F^\beta t\right]}{1 - \sum_{k=1}^{N-1} a_k\alpha_k}.$$

Example 5.3 Let the data set $\{(0, 0, 0), (2/4, -0.0327, 1.3814), (3/4, -0.2444, 2.9041), (1, -0.1497, 4.6868)\}$ be given. The hidden variable fractal interpolation function g is determined by the IFS consisting the following functions:

$$\tilde{R}_1^1(x,y,z) = \frac{1}{4}y + \frac{1}{6}z - \frac{x^2}{6(\Gamma(\beta+1))^2} - \frac{x}{2\Gamma(\beta+1)},$$
$$\tilde{R}_2^1(x,y,z) = \frac{1}{8}y + \frac{1}{12}z - \frac{11x^2}{24(\Gamma(\beta+1))^2} - 0.0327,$$
$$\tilde{R}_3^1(x,y,z) = \frac{1}{8}y + \frac{1}{12}z - \frac{x^2}{6(\Gamma(\beta+1))^2} - \frac{x}{2\Gamma(\beta+1)} - 0.2444,$$
$$\tilde{R}_1^2(x,z) = \frac{1}{4}z + \frac{x^2}{2(\Gamma(\beta+1))^2} + \frac{3x}{4\Gamma(\beta+1)},$$
$$\tilde{R}_2^2(x,z) = \frac{1}{8}z + \frac{5x^2}{8(\Gamma(\beta+1))^2} + \frac{x}{8\Gamma(\beta+1)} + 1.3814,$$
$$\tilde{R}_3^2(x,z) = \frac{1}{4}z - \frac{x^2}{2(\Gamma(\beta+1))^2} + \frac{13x}{8\Gamma(\beta+1)} + 2.9041.$$

In connection with Theorem 5.1, let us define the fractal integral of order β of $\mathbf{g} = (g_1, g_2)$ as follow

$$\tilde{g}_1(x) = \tilde{y}_N - \int_x^{x_N} S_F^\beta(g_1(t))d_F^\beta t \tag{5.27}$$

$$\tilde{g}_2(x) = \tilde{z}_N - \int_x^{x_N} S_F^\beta(g_2(t))d_F^\beta t \tag{5.28}$$

with predefined $\tilde{y}_N, \tilde{z}_N$.

Theorem 5.2 *Let* $\mathbf{g}$ *be the hidden variable fractal interpolation function generated by the IFS provided in Eq. (5.9). For given* $\tilde{y}_N, \tilde{z}_N$*, fractal integral of* $\mathbf{g}$ *is defined in Eq. (5.27), Eq. (5.28), denoted as* $\tilde{\mathbf{g}}$*. Then* $\tilde{\mathbf{g}} = (\tilde{g}_1, \tilde{g}_2)$ *is also hidden variable fractal interpolation function associated with* $\{(L_i(x), \tilde{R}_i(x,y,z))\} : i \in \mathbb{N}_{N-1}\}$*, where, for* $i \in \mathbb{N}_{N-1}$

$$\tilde{R}_i = \tilde{A}_i(y,z)^t + (\tilde{p}_i(x), \tilde{q}_i(x))^t$$
$$\tilde{A}_i = a_i \begin{bmatrix} \alpha_i & \beta_i \\ 0 & \gamma_i \end{bmatrix}$$
$$\tilde{p}_i(x) = \tilde{y}_{i+1} - a_i\alpha_i\tilde{y}_N - a_i\beta_i\tilde{z}_N - a_i\int_x^{x_N} S_F^\beta(p_i(t))d_F^\beta t,$$
$$\tilde{y}_i = \tilde{y}_N - \sum_{k=i}^{N-1} a_k\left[\alpha_k(\tilde{y}_N - \tilde{y}_1) + \beta_k(\tilde{z}_N - \tilde{z}_1) + \int_{x_1}^{x_N} S_F^\beta(p_i(t))d_F^\beta t\right],$$
$$\tilde{y}_1 = \tilde{y}_N - \frac{\sum_{k=1}^{N-1} a_k\left[\beta_k(\tilde{z}_N - \tilde{z}_1) + \int_{x_1}^{x_N} S_F^\beta(p_i(t))d_F^\beta t\right]}{1 - \sum_{k=1}^{N-1} a_k\alpha_k},$$
$$\tilde{q}_i(x) = \tilde{z}_{i+1} - a_i\gamma_i\tilde{z}_N - a_i\int_x^{x_N} S_F^\beta(q_i(t))d_F^\beta t,$$

$$\tilde{z}_i = \tilde{z}_N - \sum_{k=i}^{N-1} a_k \left[\gamma_k(\tilde{z}_N - \tilde{z}_1) + \int_{x_1}^{x_N} S_F^\beta(q_i(t)) d_F^\beta t \right],$$
$$\tilde{z}_1 = \tilde{z}_N - \frac{\sum_{k=1}^{N-1} a_k \int_{x_1}^{x_N} S_F^\beta(q_k(t)) d_F^\beta t}{1 - \sum_{k=1}^{N-1} a_k \gamma_k}.$$

Proof: Applying the definition of the fractal integral on the function g_1 gives

$$\begin{aligned}
\tilde{g}_1(L_i(x)) &= \tilde{y}_N - \int_{L_i(x)}^{x_N} S_F^\beta(g_1(t)) d_F^\beta t \\
&= \tilde{y}_N - \int_{L_i(x)}^{x_{i+1}} S_F^\beta(g_1(t)) d_F^\beta t - \int_{x_{i+1}}^{x_N} S_F^\beta(g_1(t)) d_F^\beta t \\
&= \tilde{y}_{i+1} - a_i \int_x^{x_N} S_F^\beta(g_1(L_i(t))) d_F^\beta t.
\end{aligned}$$

Since, the component of the hidden variable fractal function g_1 satisfies the functional equation given in Eq. (5.11)

$$\begin{aligned}
\tilde{g}_1(L_i(x)) &= \tilde{y}_{i+1} - a_i \int_x^{x_N} S_F^\beta(\alpha_i g_1(t) + \beta_i g_2(t) + p_i(t)) d_F^\beta t \\
&= \tilde{y}_{i+1} - a_i\alpha_i \int_x^{x_N} S_F^\beta(g_1(t)) d_F^\beta t - a_i\beta_i \int_x^{x_N} S_F^\beta(g_2(t)) d_F^\beta t - a_i \int_x^{x_N} S_F^\beta(p_i(t)) d_F^\beta t.
\end{aligned}$$

Equation (5.27) and Eq. (5.28) provide

$$\begin{aligned}
\tilde{g}_1(L_i(x)) &= \tilde{y}_{i+1} - a_i\alpha_i(\tilde{y}_N - \tilde{g}_1(x)) - a_i\beta_i(\tilde{z}_N - \tilde{g}_2(x)) - a_i \int_x^{x_N} S_F^\beta(p_i(t)) d_F^\beta t \\
&= a_i\alpha_i \tilde{g}_1(x) + a_i\beta_i \tilde{g}_2(x) + \tilde{p}_i(x),
\end{aligned}$$

where $\tilde{p}_i(x) = \tilde{y}_{i+1} - a_i\alpha_i\tilde{y}_N - a_i\beta_i\tilde{z}_N - a_i \int_x^{x_N} S_F^\beta(p_i(t)) d_F^\beta t$. Similarly,

$$\begin{aligned}
\tilde{g}_2(L_i(x)) &= \tilde{z}_N - \int_{L_i(x)}^{x_N} S_F^\beta(g_2(t)) d_F^\beta t \\
&= \tilde{z}_N - \int_{L_i(x)}^{x_{i+1}} S_F^\beta(g_2(t)) d_F^\beta t - \int_{x_{i+1}}^{x_N} S_F^\beta(g_2(t)) d_F^\beta t \\
&= \tilde{z}_{i+1} - a_i \int_x^{x_N} S_F^\beta(g_2(L_i(t))) d_F^\beta t.
\end{aligned}$$

Since, g_2 is component of hidden variable fractal function and satisfies the functional equation given in Eq. (5.11)

$$\begin{aligned}
\tilde{g}_2(L_i(x)) &= \tilde{z}_{i+1} - a_i \int_x^{x_N} S_F^\beta(\gamma_i g_2(t) + q_i(t)) d_F^\beta t \\
&= \tilde{z}_{i+1} - a_i\gamma_i \int_x^{x_N} S_F^\beta(g_2(t)) d_F^\beta t - a_i \int_x^{x_N} S_F^\beta(q_i(t)) d_F^\beta t.
\end{aligned}$$

Equation (5.28) gives

$$\begin{aligned}\tilde{g}_2(L_i(x)) &= \tilde{z}_{i+1} - a_i\gamma_i(\tilde{z}_N - \tilde{g}_2(x)) - a_i\int_x^{x_N} S_F^\beta(q_i(t))d_F^\beta t.\\ &= a_i\gamma_i\tilde{g}_2(x) + \tilde{q}_i(x),\end{aligned}$$

where $\tilde{q}_i(x) = \tilde{z}_{i+1} - a_i\gamma_i\tilde{z}_N - a_i\int_x^{x_N} S_F^\beta(q_i(t))d_F^\beta t$. Hence, $\tilde{\mathbf{g}} = (\tilde{g}_1, \tilde{g}_2)$ is hidden variable fractal interpolation function generated by the IFS $\{(L_i(x), \tilde{R}_i(x,y,z)) : i \in \mathbb{N}_{N-1}\}$. Take $x = x_1, L_i(x_1) = x_i$, then

$$\begin{aligned}\tilde{y}_i &= \tilde{y}_{i+1} - a_i\alpha_i(\tilde{y}_N - \tilde{y}_1) - a_i\beta_i(\tilde{z}_N - \tilde{z}_1) - a_i\int_{x_1}^{x_N} S_F^\beta(p_i(t))d_F^\beta t\\ \tilde{y}_{i+1} - \tilde{y}_i &= a_i\left[\alpha_i(\tilde{y}_N - \tilde{y}_1) + \beta_i(\tilde{z}_N - \tilde{z}_1) + \int_{x_1}^{x_N} S_F^\beta(p_i(t))d_F^\beta t\right]\\ \tilde{z}_i &= \tilde{z}_{i+1} - a_i\gamma_i(\tilde{z}_N - \tilde{z}_1) - a_i\int_{x_1}^{x_N} S_F^\beta(q_i(t))d_F^\beta t\\ \tilde{z}_{i+1} - \tilde{z}_i &= a_i\gamma_i(\tilde{z}_N - \tilde{z}_1) + a_i\int_{x_1}^{x_N} S_F^\beta(q_i(t))d_F^\beta t.\end{aligned}$$

Clearly, $\tilde{y}_i = \tilde{y}_N - \sum_{k=i}^{N=1}(\tilde{y}_{k+1} - \tilde{y}_k)$, this equation yields the new data points

$$\tilde{y}_i = \tilde{y}_N - \sum_{k=i}^{N-1} a_k\left[\alpha_k(\tilde{y}_N - \tilde{y}_1) + \beta_k(\tilde{z}_N - \tilde{z}_1) + \int_{x_1}^{x_N} S_F^\beta(p_k(t))d_F^\beta t\right]. \tag{5.29}$$

$$\tilde{z}_i = \tilde{z}_N - \sum_{k=i}^{N-1} a_k\left[\gamma_k(\tilde{z}_N - \tilde{z}_1) + \int_{x_1}^{x_N} S_F^\beta(p_k(t))d_F^\beta t\right]. \tag{5.30}$$

Taking $i = 1$ in Eq. (5.29) and Eq. (5.30) give the endpoints $\tilde{y}_1, \tilde{z}_1$ of the new data set

$$\begin{aligned}\tilde{y}_1 &= \tilde{y}_N - \sum_{k=1}^{N-1} a_k\left[\alpha_k(\tilde{y}_N - \tilde{y}_1) + \beta_k(\tilde{z}_N - \tilde{z}_1) + \int_{x_1}^{x_N} S_F^\beta(p_k(t))d_F^\beta t\right]\\ (\tilde{y}_N - \tilde{y}_1)\left[1 - \sum_{k=1}^{N-1} a_k\alpha_k\right] &= \sum_{k=1}^{N-1} a_k\left[\beta_k(\tilde{z}_N - \tilde{z}_1) + \int_{x_1}^{x_N} S_F^\beta(p_k(t))d_F^\beta t\right].\end{aligned}$$

A similar argument gives the endpoint $\tilde{z}_1$. Thus, $\tilde{\mathbf{g}} = (\tilde{g}_1, \tilde{g}_2)$ is hidden variable fractal interpolation function which interpolates the new set of data $\{(x_i, \tilde{y}_i, \tilde{z}_i) : i \in \mathbb{N}_N\}$, where, for each $i \in \mathbb{N}_N\backslash\{1\}$,

$$\tilde{y}_i = \tilde{y}_N - \sum_{k=i}^{N-1} a_k \left[\alpha_k(\tilde{y}_N - \tilde{y}_1) + \beta_k(\tilde{z}_N - \tilde{z}_1) + \int_{x_1}^{x_N} S_F^\beta(p_i(t))d_F^\beta t\right],$$

$$\tilde{z}_i = \tilde{z}_N - \sum_{k=i}^{N-1} a_k \left[\gamma_k(\tilde{z}_N - \tilde{z}_1) + \int_{x_1}^{x_N} S_F^\beta(q_i(t))d_F^\beta t\right],$$

$$\tilde{y}_1 = \tilde{y}_N - \frac{\sum_{k=1}^{N-1} a_k \left[\beta_k(\tilde{z}_N - \tilde{z}_1) + \int_{x_1}^{x_N} S_F^\beta(p_i(t))d_F^\beta t\right]}{1 - \sum_{k=1}^{N-1} a_k\alpha_k},$$

$$\tilde{z}_1 = \tilde{z}_N - \frac{\sum_{k=1}^{N-1} a_k \int_{x_1}^{x_N} S_F^\beta(q_k(t))d_F^\beta t}{1 - \sum_{k=1}^{N-1} a_k\gamma_k}.$$

Theorems 5.1 and 5.2 provide the fractal calculus of the hidden variable fractal interpolation function with predefined endpoint. Let us examine the fractal calculus of α-fractal function, define

$$\tilde{g}^\alpha(x) = \tilde{y}_1 + \int_{x_1}^{x} S_F^\beta(g^\alpha(t))d_F^\beta t \tag{5.31}$$

where $\tilde{y}_1$ is predefined.

Theorem 5.3 *If g^α is the α-fractal interpolation function with $\{(L_i(x), R_i(x,y)) : i \in \mathbb{N}_{N-1}\}$ where R_i is given in Eq. (5.13) and for a given $\tilde{y}_1$, fractal integral of g^α is defined in Eq. (5.31). Suppose the functions $S_F^\beta F$ and $S_F^\beta B$ are defined as the fractal integrals of f and b, $S_F^\beta F(x) = \int_{x_1}^{x} S_F^\beta(f(t))d_F^\beta t$, $S_F^\beta B(x) = \int_{x_1}^{x} S_F^\beta(b(t))d_F^\beta t$ with $\tilde{y}_1 = S_F^\beta F(x_1) = S_F^\beta B(x_1) = 0$ and $S_F^\beta F(x_N) = S_F^\beta B(x_N)$ then $\tilde{g}^\alpha$ is also α-fractal interpolation function associated with $\{(L_i(x), \tilde{R}_i(x,y)) : i \in \mathbb{N}_{N-1}\}$, where, for $i \in \mathbb{N}_{N-1}$,*

$$\begin{aligned}
\tilde{R}_i &= a_i\alpha_i y + \tilde{q}_i(x), \\
a_i &= \frac{x_{i+1} - x_i}{x_N - x_1}, \\
\tilde{q}_i(x) &= S_F^\beta(F \circ L_i)(x) - a_i\alpha_i S_F^\beta B(x), \\
\tilde{y}_{i+1} &= a_i\alpha_i\tilde{y}_N + S_F^\beta F \circ L_i(x_N) - a_i\alpha_i S_F^\beta B(x_N), \\
\tilde{y}_N &= \frac{S_F^\beta F \circ L_{N-1}(x_N) - a_{N-1}\alpha_{N-1} S_F^\beta B(x_N)}{1 - a_{N-1}\alpha_{N-1}}.
\end{aligned}$$

Proof: Applying the definition of fractal integral on the g^α function gives

$$\begin{aligned}
\tilde{g}^\alpha(L_i(x)) &= \tilde{y}_1 + \int_{x_1}^{L_i(x)} S_F^\beta(g^\alpha(t))d_F^\beta t \\
&= \tilde{y}_1 + \int_{x_1}^{x_i} S_F^\beta(g^\alpha(t))d_F^\beta t + \int_{x_i}^{L_i(x)} S_F^\beta(g^\alpha(t))d_F^\beta t \\
&= \tilde{y}_i + a_i \int_{x_1}^{x} S_F^\beta(g^\alpha(L_i(t)))d_F^\beta t.
\end{aligned}$$

Since, α-fractal function g^α satisfies the functional equation given in Eq. (5.14)

$$\begin{aligned}\tilde{g}^\alpha(L_i(x)) &= \tilde{y}_i + a_i \int_{x_1}^{x} S_F^\beta(f \circ L_i(t) + \alpha_i(g^\alpha(t) - b(t)))d_F^\beta t \\ &= \tilde{y}_i + a_i\alpha_i \int_{x_1}^{x} S_F^\beta(g^\alpha(t))d_F^\beta t + a_i \int_{x_1}^{x} S_F^\beta(f \circ L_i(t))d_F^\beta t - a_i\alpha_i \int_{x_1}^{x} S_F^\beta(b(t))d_F^\beta t.\end{aligned}$$

By Eq. (5.31) one can get

$$\begin{aligned}\tilde{g}^\alpha(L_i(x)) &= \tilde{y}_i + a_i\alpha_i(\tilde{g}^\alpha(x) - \tilde{y}_1) + a_i \int_{x_1}^{x} S_F^\beta(f \circ L_i(t))d_F^\beta t - a_i\alpha_i \int_{x_1}^{x} S_F^\beta(b(t))d_F^\beta t \\ &= \tilde{y}_i + a_i\alpha_i\tilde{g}^\alpha(x) + \int_{x_1}^{x} S_F^\beta(F \circ L_i)'(u)d_F^\beta u - a_i\alpha_i S_F^\beta B(x) \\ &= \tilde{y}_i + a_i\alpha_i\tilde{g}^\alpha(x) + S_F^\beta(F \circ L_i)(x) - S_F^\beta(F \circ L_i)(x_1) - a_i\alpha_i S_F^\beta B(x) \\ &= a_i\alpha_i\tilde{g}^\alpha(x) + S_F^\beta(F \circ L_i)(x) - a_i\alpha_i S_F^\beta B(x) \\ &= a_i\alpha_i\tilde{g}^\alpha(x) + \tilde{q}_i(x),\end{aligned}$$

where $\tilde{q}_i(x) = S_F^\beta(F \circ L_i)(x) - a_i\alpha_i S_F^\beta B(x)$. Now, it is easy to verify that

$$\begin{aligned}\tilde{q}_i(x_0) &= S_F^\beta(F \circ L_i)(x_1) - a_i\alpha_i S_F^\beta B(x_0) \\ &= \tilde{y}_i - a_i\alpha_i\tilde{y}_1, \\ \tilde{q}_i(x_N) &= S_F^\beta(F \circ L_i)(x_N) - a_i\alpha_i S_F^\beta B(x_N) \\ &= \tilde{y}_{i+1} - a_i\alpha_i\tilde{y}_N,\end{aligned}$$

which clearly implies $\tilde{R}_i(x_1, \tilde{y}_1) = \tilde{y}_i$ and $\tilde{R}_i(x_N, \tilde{y}_N) = \tilde{y}_{i+1}$. As $\tilde{g}^\alpha$ satisfies the continuity conditions at the joints, it is also a α-fractal interpolation function generated by the IFS $\{(L_i(x), \tilde{R}_i(x, y)) : i \in \mathbb{N}_{N-1}\}$. To find the new set of data points of the interpolant $\tilde{g}^\alpha$, take $x = x_N,\ L_i(x_N) = x_{i+1}$ in the following equation,

$$\tilde{g}^\alpha(L_i(x)) = a_i\alpha_i\tilde{g}^\alpha(x) + S_F^\beta(F \circ L_i)(x) - a_i\alpha_i S_F^\beta B(x).$$

then

$$\tilde{y}_{i+1} = a_i\alpha_i\tilde{y}_N + S_F^\beta F \circ L_i(x_N) - a_i\alpha_i S_F^\beta B(x_N).$$

Substituting $i = N - 1$ in the above equation, one can get the end point $\tilde{y}_N$ as,

$$\tilde{y}_N = \frac{S_F^\beta F \circ L_{N-1}(x_N) - a_{N-1}\alpha_{N-1} S_F^\beta B(x_N)}{1 - a_{N-1}\alpha_{N-1}}.$$

Theorem 5.3 assurances the fractal integral of a α-fractal function is also α-fractal function which interpolates the new set of data points $\{(x_i, \tilde{y}_i) : i \in \mathbb{N}_N\}$. In connection with Theorem 1.3 define

$$\tilde{g}^\alpha(x) = \tilde{y}_N - \int_{x}^{x_N} S_F^\beta(g^\alpha(t))d_F^\beta t \tag{5.32}$$

with predefined $\tilde{y}_N$.

Theorem 5.4 *If g^α is the α-fractal interpolation function with $\{(L_i(x), R_i(x,y)) : i \in \mathbb{N}_{N-1}\}$ where R_i is given in Eq. (5.13) and, for predefined $\tilde{y}_N$, fractal integral of g^α is given in Eq. (5.32). Suppose the functions $S_F^\beta F$ and $S_F^\beta B$ are defined as the fractal integrals of f and b, $S_F^\beta F(x) = -\int_x^{x_N} S_F^\beta(f(t))d_F^\beta t$, $S_F^\beta B(x) = -\int_x^{x_N} S_F^\beta(b(t))d_F^\beta t$ with $S_F^\beta F(x_1) = S_F^\beta B(x_1)$ and $\tilde{y}_N = S_F^\beta F(x_N) = S_F^\beta B(x_N) = 0$ then $\tilde{g}^\alpha$ is also a α-fractal interpolation function associated with $\{(L_i(x), \tilde{R}_i(x,y)) : i \in \mathbb{N}_{N-1}\}$, where, for $i \in \mathbb{N}_{N-1}$,*

$$\begin{aligned}
\tilde{R}_i &= a_i\alpha_i y + \tilde{q}_i(x),\\
a_i &= \frac{x_{i+1} - x_i}{x_N - x_1},\\
\tilde{q}_i(x) &= S_F^\beta(F \circ L_i)(x) - a_i\alpha_i S_F^\beta B(x),\\
\tilde{y}_i &= a_i\alpha_i\tilde{y}_1 + S_F^\beta F \circ L_i(x_1) - a_i\alpha_i S_F^\beta B(x_1),\\
\tilde{y}_1 &= \frac{S_F^\beta F \circ L_1(x_1) - a_1\alpha_1 S_F^\beta B(x_1)}{1 - a_1\alpha_1}.
\end{aligned}$$

Proof: Applying the definition of the fractal integral of order β on the g^α function gives

$$\begin{aligned}
\tilde{g}^\alpha(L_i(x)) &= \tilde{y}_N - \int_{L_i(x)}^{x_N} S_F^\beta(g^\alpha(t))d_F^\beta t\\
&= \tilde{y}_N - \int_{L_i(x)}^{x_{i+1}} S_F^\beta(g^\alpha(t))d_F^\beta t - \int_{x_{i+1}}^{x_N} S_F^\beta(g^\alpha(t))d_F^\beta t\\
&= \tilde{y}_{i+1} - a_i \int_x^{x_N} S_F^\beta(g^\alpha(L_i(t)))d_F^\beta t.
\end{aligned}$$

Since, α-fractal function g^α satisfies the functional equation given in Eq. (5.14)

$$\begin{aligned}
\tilde{g}^\alpha(L_i(x)) &= \tilde{y}_{i+1} - a_i \int_x^{x_N} S_F^\beta(f \circ L_i(t) + \alpha_i(g^\alpha(t) - b(t)))d_F^\beta t\\
&= \tilde{y}_{i+1} - a_i\alpha_i \int_x^{x_N} S_F^\beta(g^\alpha(t))d_F^\beta t - a_i \int_x^{x_N} S_F^\beta(f \circ L_i(t))d_F^\beta t + a_i\alpha_i \int_x^{x_N} S_F^\beta(b(t))d_F^\beta t.
\end{aligned}$$

Equation (5.31) yields

$$\begin{aligned}
\tilde{g}^\alpha(L_i(x)) &= \tilde{y}_{i+1} + a_i\alpha_i(\tilde{g}^\alpha(x) - \tilde{y}_N) - a_i \int_x^{x_N} S_F^\beta(f \circ L_i(t))d_F^\beta t + a_i\alpha_i \int_x^{x_N} S_F^\beta(b(t))d_F^\beta t\\
&= \tilde{y}_{i+1} + a_i\alpha_i\tilde{g}^\alpha(x) - \int_x^{x_N} S_F^\beta(F \circ L_i)'(u)d_F^\beta u - a_i\alpha_i S_F^\beta B(x)\\
&= \tilde{y}_{i+1} + a_i\alpha_i\tilde{g}^\alpha(x) - S_F^\beta(F \circ L_i)(x_N) + S_F^\beta(F \circ L_i)(x) - a_i\alpha_i S_F^\beta B(x)\\
&= a_i\alpha_i\tilde{g}^\alpha(x) + S_F^\beta(F \circ L_i)(x) - a_i\alpha_i S_F^\beta B(x)\\
&= a_i\alpha_i\tilde{g}^\alpha(x) + \tilde{q}_i(x),
\end{aligned}$$

where $\tilde{q}_i(x) = S_F^\beta(F \circ L_i)(x) - a_i\alpha_i S_F^\beta B(x)$. Now, it is easy to verify that

$$\begin{aligned}\tilde{q}_i(x_0) &= S_F^\beta(F \circ L_i)(x_1) - a_i\alpha_i S_F^\beta B(x_0)\\ &= \tilde{y}_i - a_i\alpha_i\tilde{y}_1,\\ \tilde{q}_i(x_N) &= S_F^\beta(F \circ L_i)(x_N) - a_i\alpha_i S_F^\beta B(x_N)\\ &= \tilde{y}_{i+1} - a_i\alpha_i\tilde{y}_N,\end{aligned}$$

which clearly implies $\tilde{R}_i(x_1, \tilde{y}_1) = \tilde{y}_i$ and $\tilde{R}_i(x_N, \tilde{y}_N) = \tilde{y}_{i+1}$. As $\tilde{g}^\alpha$ satisfies the continuity conditions at the joints, it is also a α-fractal interpolation function generated by the IFS $\{(L_i(x), \tilde{R}_i(x, y)) : i \in \mathbb{N}_{N-1}\}$. To find the new set of data points of the interpolant $\tilde{g}^\alpha$, take $x = x_1,\ L_i(x_1) = x_i$ in the following equation,

$$\tilde{g}^\alpha(L_i(x)) = a_i\alpha_i\tilde{g}^\alpha(x) + S_F^\beta(F \circ L_i)(x) - a_i\alpha_i S_F^\beta B(x).$$

then

$$\tilde{y}_i = a_i\alpha_i\tilde{y}_1 + S_F^\beta F \circ L_i(x_1) - a_i\alpha_i S_F^\beta B(x_1).$$

Substituting $i = 1$ in the above equation, one can get the end point $\hat{y}_N$ as,

$$\tilde{y}_1 = \frac{S_F^\beta F \circ L_1(x_1) - a_1\alpha_1 S_F^\beta B(x_1)}{1 - a_1\alpha_1}.$$

5.6 Concluding Remarks

In recent days, the analysis of fractal function by fractional calculus is getting attention in the field of non-linear analysis. The mass function is associated with the process of finding fractal integral and the fractal derivative coincides with classical calculus as the constant function which has zero derivatives which contrasted with the fractional derivative. Hence, this article contributes to incorporate the notions of fractal interpolation functions and fractal calculus. This amalgam is performed by discussing the fractal integral of hidden variable fractal interpolation function with the initial conditions. Further, the fractal integral of α-fractal interpolation function is investigated with the predefined initial condition at the endpoints. The present study suggests some other avenues to analyse the local derivative of the fractal interpolation function.

References

[1] Barnsley, M.F. 1986. Fractal functions and interpolation. Constr. Approx., 2(1): 303–329.
[2] Navascués, M.A. 2005. Fractal polynomial interpolation. Z. Anal. Anwend., 25(2): 401–418.
[3] Barnsley, M.F., J. Elton, D. Hardin and P.R. Massopust. 1989. Hidden variable fractal interpolation functions. SIAM J. Math. Anal., 20(5):1218–1242.
[4] Bouboulis, P. and L. Dalla. 2005. Hidden variable vector valued fractal interpolation functions. Fractals, 13(3): 227–232.
[5] Chand, A.K.B. and G.P. Kappor. 2006. Spline coalescence hidden variable fractal interpolation function. J. Appl. Math. Article ID 36829: 1–17.
[6] Chand, A.K.B. and G.P. Kappor. 2007. Smoothness analysis of coalescence hidden variable fractal interpolation functions. Int. J. Nonlinear Sci., 15–26.
[7] Massopust, P.R. 1994. Fractal Functions. Fractal Surfaces and Wavelets. Academic Press.

[8] Banerjee, S., D. Easwaramoorthy and A. Gowrisankar. 2021. Fractal Functions, Dimensions and Signal Analysis, Springer, Cham.
[9] Banerjee, S., M.K. Hassan, S. Mukherjee and A. Gowrisankar. 2019. Fractal Patterns in Nonlinear Dynamics and Applications, CRC Press, Boca Raton.
[10] Parvate, A. and A.D. Gangal. 2009. Calculus on fractal subsets of real line–I: Formulation. Fractals, 17(1): 53–81.
[11] Parvate, A., S. Satin and A.D. Gangal. 2011. Calculus on fractal curves in $\mathbb{R}^n$. Fractals, 19(1): 15–27.
[12] Parvate, A. and A.D. Gangal. 2011. Calculus on fractal subsets of real line–II: Conjugacy with ordinary calculus. Fractals, 19(3): 271–290.
[13] Golmankhaneh, A.K. 2017. On the calculus of parameterized fractal curves. Turk. J. Phys., 41(5): 418–425.
[14] Golmankhaneh, A.K. and C. Tunç. 2019. Stochastic differential equations on fractal sets. Stochastics, 1–17.
[15] Golmankhaneh, A.K. and C. Tunç. 2019. Sumudu transform in fractal calculus. Appl. Math. Comput., 350: 386–401.
[16] Tatom, F.B. 1995. The relationship between fractional calculus and fractal. Fractals, 3(1):217–229.
[17] Yao, K., W.Y. Su and S.P. Zhou. 2005. On the connection between the order of fractional calculus and the dimensions of a fractal function. Chaos Solitons Fractals, 23: 621–629.
[18] Ruan, H.J., W.Y. Su and K. Yao. 2009. Box dimension and fractional integral of linear fractal interpolation functions. J. Approx. Theory, 161: 187–197.
[19] He, J.H. 2018. Fractal calculus and its geometrical explanation. Results Phys., 10: 272–276.
[20] Chen, W. and Y. Liang. 2017. New methodologies in fractional and fractal derivatives modelling. Chaos Solitons Fractals, 102: 72–77.
[21] Liang, Y.S. and W.Y. Su. 2016. Fractal dimensions of fractional integral of continuous functions. Acta Math. Sin., 32(12): 1494–1508.
[22] Liang, Y.S. and Q. Zhang. 2016. A type of fractal interpolation functions and their fractional calculus. Fractals, 24(2): 1650026.
[23] Gowrisankar, A. and R. Uthayakumar. 2016. Fractional calculus on fractal interpolation function for a sequence of data with countable iterated function system. Mediterr. J. Math., 13(6): 3887–3906.
[24] Wang, H.Y. and J.S. Yu. 2013. Fractal interpolation functions with variable parameters and their analytical properties. J. Approx. Theory, 175: 1–8.
[25] Xiao, E.W. and H.D. Jun. 2017. Box dimension of Hadamard fractional integral of continuous functions of bounded and unbounded variation. Fractals, 25(3): 1750035.
[26] Li, Y. and Y. Liang. 2017. Upper bound estimation of fractal dimension of fractional calculus of continuous functions. Advances in Analysis 2(2): 121–128.
[27] Gowrisankar, A. and M.G.P. Prasad. 2019. Riemann-Liouville calculus on quadratic fractal interpolation function with variable scaling factors. The Journal of Analysis, 27(2): 347–363.
[28] Priyanka, T.M.C. and A. Gowrisankar. 2021. Analysis on Weyl-Marchaud fractional derivative for types of fractal interpolation function with fractal dimension. Fractals, doi.org/10.1142/S0218348X21502157.
[29] Prasad, P.K., A. Gowrisankar, A. Saha and S. Banerjee. 2020. Dynamical properties and fractal patterns of nonlinear waves in solar wind plasma. Phys. Scr., 95(6): 065603.
[30] Fataf, N.A.A., A. Gowrisankar and S. Banerjee. 2020. In search of self-similar chaotic attractors based on fractal function with variable scaling approximately. Phys. Scr., 95(7): 075206.
[31] Liang, Y.S. 2019. Progress on estimation of fractal dimensions of fractional calculus of continuous functions. Fractals, 27(5): 19500849.

Chapter 6

On the Borel Regularity of the Relative Centered Multifractal Measures

Zied Douzi and Bilel Selmi*

6.1 Introduction

The Hausdorff measure was introduced in more general form at the beginning of the last century by Carathéodory [7] and Hausdorff [20]. The spherical measure is also just a specific case of the initial definition. It is very hard to overestimate the importance of the Hausdorff measure in geometric measure theory. But besides playing a central role in this area, it is also of great importance to a wide range of other mathematical areas. For further applications and a more general definition, one should consult the book on Hausdorff measures [26]. The centered Hausdorff and packing measures were introduced by Tricot et al. in [25, 32] for general continuous and doubling gauge functions depending only on the radius of the ball and by Olsen in [23]. Later, these measures were generalized by Cole in [8] and Peyrière in [24] as defined in Section 2 which is analogous to Billingsley's classical Hausdorff measure in [5, 6]. The relative multifractal dimensions b and B defined by these measures were used to give estimates for the multifractal spectrum of a measure. In several recent papers on multifractal analysis, and this type of multifractal analysis has re-emerged as mathematicians and physicists have begun to discuss the idea of performing multifractal analysis with respect to an arbitrary reference measure, see for example [2, 11–15, 28, 29, 16]. In [21], the authors prove a modification of this type of analysis. Instead of studying sets of points with a local dimension which is given with respect to Lebesgue measure, they studied sets of points with a local dimension given with respect to a non-atomic probability measure ν and checks an auxiliary condition. It is very natural to study this formalism of multifractal analysis for what differs slightly from what was introduced in [8]. The difference between the two types is that we used centered ν-δ-coverings and centered ν-δ-packings rather than centered δ-coverings and δ-pakings. These relative multifractal measures and dimensions have been used for other purposes as well, for example, [1, 2, 3, 4, 30, 31] and have recently become an object of study themselves, see [9, 10, 17].

An interesting open question about the centered Hausdorff measure is whether it is a Borel regular measure. This problem has been asked several times during the 90's by many authors, see for example in [18, 22, 23]. This question has recently been answered affirmatively Schechter in [27]. In fact, Schechter proved that if μ is a Radon measure on a metric space satisfying the doubling condition, then the centered (multifractal) Hausdorff measure is Borel regular. Moreover, he also constructed a Radon measure μ in the plane not satisfying the doubling condition for which the centered Hausdorff measure is not Borel regular. This is the idea that we refine to get our results. Another interesting question, which has been asked several times about the centered relative multifractal measures, is whether it is Borel regular. A positive answer is

Analysis, Probability & Fractals Laboratory LR18ES17, Department of Mathematics, Faculty of Sciences of Monastir, University of Monastir, 5000-Monastir, Tunisia.
Email: zied.douzi@fsm.rnu.tn
* Corresponding author: bilel.selmi@fsm.rnu.tn,bilel.selmi@isetgb.rnu.tn

given, using the equivalence of the spherical measure and the centered relative multifractal measure for all gauge functions.

This chapter shows that the relative packing measure $\mathscr{P}^g$ is Borel regular, and the usual spherical measure $\mathscr{U}^g$ and the relative Hausdorff measure $\mathscr{H}^g$ are equivalent in the case where μ and ν satisfy the doubling condition. Also, this chapter shows that this statement is true even without the doubling condition if $q \geq 1$ and $t > 0$ or if $q \leq 0$ and $t \leq 0$ which implies, in particular, $\mathscr{H}^g$ is a Borel regular measure. In addition, there exist two measures not satisfying the doubling condition for which the centered relative multifractal Hausdorff measure is not Borel regular for $0 < q < 1$.

6.2 The Relative Multifractal Measures

The main purpose of this section is to review briefly some known concepts that will be applied next in the development of our main results, and which will enlighten in some sense the difference with existing works, and the novelty in the present work. Readers maybe referred essentially to [21] for more details. We deal with a metric space (X, d) having the Besicovitch property (or for which the Besicovitch covering theorem holds), see [22]: There exists an integer constant ξ such that one can extract ξ countable families $\{\{B_{j,k}\}_k\}_{1\leq j\leq\xi}$ from any collection $\mathcal{B}$ of balls so that

1. $\bigcup_{j,k} B_{j,k}$ contains the centers of the elements of $\mathcal{B}$,
2. for any j and $k \neq k'$, $B_{j,k} \cap B_{j,k'} = \emptyset$.

Let (X, d) be a separable metric space for which the Besicovitch covering theorem holds, $E \subset X$ and μ, ν be two probability measures on X. Throughout this chapter, $B(x, r)$ stands for the closed ball

$$B(x,r) = \{y \in X \mid \, d(x,y) \leq r\}.$$

For $q, t \in \mathbb{R}$, let

$$g(x,r) = \mu(B(x,r))^q \nu(B(x,r))^t.$$

Definition 6.1 [21]Let $\delta > 0$.

1. A countable family $\Big(B(x_i, r_i)\Big)_i$ of closed balls in X is called

   a. a centered ν-δ-packing of E if

   $$\forall i,\ x_i \in E,\ 0 < \nu\left(B(x_i,r_i)\right) < \delta \quad \text{and} \quad B(x_i,r_i) \cap B(x_j,r_j) = \emptyset, \quad \text{if} \quad i \neq j.$$

   b. a ν-δ-covering of E if

   $$\forall i,\ 0 < \nu\left(B(x_i,r_i)\right) < \delta \quad \text{and} \quad E \subset \bigcup_i B(x_i,r_i).$$

2. We call $(B(x_i, r_i))_i$ a centered ν-δ-covering of E if in addition the balls are centered in E.
3. We say that ν is non-atomic if $\nu(\{x\}) = 0$, for any $x \in X$.

Throughout this chapter, we assume that ν is a non-atomic Borel probability measure on X and μ be a probability measure on X with $S = \mathrm{supp}(\mu) \cap \mathrm{supp}(\nu) \neq \emptyset$. Let us now define the measures that are the central object of study in this chapter.

Definition 6.2 [21] Let $\delta > 0$. For $\emptyset \neq E \subset X$, we consider

$$\mathcal{U}_\delta^g(E) = \inf\left\{\sum_i g(x_i, r_i) \,\Big|\, \Big(B(x_i, r_i)\Big)_i \text{ is a } \nu - \delta - \text{covering of } E\right\},$$

and we can set $\mathcal{U}_\delta^g(\emptyset) = 0$. The spherical measure is defined as follows

$$\mathcal{U}^g(E) = \lim_{\delta \downarrow 0} \mathcal{U}_\delta^g(E) = \sup_{\delta > 0} \mathcal{U}_\delta^g(E).$$

The relative centered Hausdorff measure is defined by

$$\overline{\mathcal{H}}_\delta^g(E) = \inf\left\{\sum_i g(x_i, r_i) \,\Big|\, \Big(B(x_i, r_i)\Big)_i \text{ is a centered } \nu - \delta - \text{covering of } E\right\}, \quad E \neq \emptyset.$$

Moreover, we can set $\overline{\mathcal{H}}_\delta^g(\emptyset) = 0$. Now, let

$$\overline{\mathcal{H}}^g(E) = \lim_{\delta \downarrow 0} \overline{\mathcal{H}}_\delta^g(E) = \sup_{\delta > 0} \overline{\mathcal{H}}_\delta^g(E)$$

and

$$\mathcal{H}^g(E) = \sup_{F \subseteq E} \overline{\mathcal{H}}^g(F).$$

We also make the dual definitions for relative centered packing measure

$$\overline{\mathcal{P}}_\delta^g(E) = \inf\left\{\sum_i g(x_i, r_i) \,\Big|\, \Big(B(x_i, r_i)\Big)_i \text{ is a centered } \nu - \delta - \text{packing of } E\right\}, \quad E \neq \emptyset.$$

Moreover, we can set $\overline{\mathcal{P}}_\delta^g(\emptyset) = 0$. Now, let

$$\overline{\mathcal{P}}^g(E) = \lim_{\delta \downarrow 0} \overline{\mathcal{P}}_\delta^g(E) = \sup_{\delta > 0} \overline{\mathcal{P}}_\delta^g(E)$$

and

$$\mathcal{P}^g(E) = \inf_{E \subseteq \cup_i E_i} \sum_i \overline{\mathcal{P}}^g(E_i)$$

with the conventions $0^0 = 1$ and $0^q = \infty$ for $q < 0$.

Here we recall the definition of the Borel regularity of measures.

Definition 6.3 [27] A Radon measure μ will always be taken to be a metric outer measure; hence all Borel sets are measurable. We say that μ is Borel regular if we can find for every $E \subset X$ a Borel set $F \supset E$ with $\mu(E) = \mu(F)$.

Proposition 6.1 *Let X be a separable metric space and $E_1, E_2 \subset X$. Then*

1. $\overline{\mathcal{H}}^g(E_1 \cup E_2) \leqslant \overline{\mathcal{H}}^g(E_1) + \overline{\mathcal{H}}^g(E_2)$ *and if* $d(E_1, E_2) > 0$ *we also get*

$$\overline{\mathcal{H}}^g(E_1 \cup E_2) = \overline{\mathcal{H}}^g(E_1) + \overline{\mathcal{H}}^g(E_2).$$

2. *$\mathscr{H}^g$ and $\mathscr{P}^g$ are Borel measures.*
3. *$\mathscr{U}^g$ is a Borel regular measure.*

Proof An exhaustive proof of this proposition would require considerable repetition. For this we only prove the results for $\mathscr{H}^g$, the other assertions are similar.

1. Let's prove that $\overline{\mathscr{H}}^g$ is sub-additive. Let $E = \bigcup_{n=1}^{\infty} E_n \subset X$ and $\delta > 0$. For each n, let $\left\{B\left(x_{ni}, r_{ni}\right)\right\}_i$ be a centered ν-δ-covering of E_n such that

$$\sum_{i=1}^{\infty} g\left(x_{ni}, r_{ni}\right) \leq \overline{\mathscr{H}}^g_\delta(E_n) + \frac{\delta}{2^n}.$$

Therefore $\left\{B\left(x_{ni}, r_{ni}\right)\right\}_{i,n}$ is a centered ν-δ-covering of E and we have

$$\overline{\mathscr{H}}^g_\delta(E) \leq \sum_{n=1}^{\infty}\sum_{i=1}^{\infty} g\left(x_{ni}, r_{ni}\right) \leq \sum_{n=1}^{\infty} \overline{\mathscr{H}}^g_\delta(E_n) + \delta.$$

When $\delta \downarrow 0$, one gets

$$\overline{\mathscr{H}}^g(E) \leq \sum_{n=1}^{\infty} \overline{\mathscr{H}}^g(E_n).$$

The other inequality will be proved in the following assertion.
2. We will just prove that $\mathscr{H}^g$ is a Borel measure. The proof that $\mathscr{P}^g$ is Borel measure follows with a similar arguments.

(i)- Let us prove that $\mathscr{H}^g$ is monotone. Let $E \subset F \subset X$, then

$$\mathscr{H}^g(E) = \sup_{A \subset E} \overline{\mathscr{H}}^g(A) \leq \sup_{A \subset F} \overline{\mathscr{H}}^g(A) = \mathscr{H}^g(F).$$

(ii)- We prove that $\mathscr{H}^g$ is sub-additive. Let $E = \bigcup_{n=1}^{\infty} E_n \subset X$ and $A \subset E$. One has

$$\overline{\mathscr{H}}^g(A) \leq \sum_{n=1}^{\infty} \overline{\mathscr{H}}^g(E_n \cap A) \leq \sum_{n=1}^{\infty} \mathscr{H}^g(E_n),$$

which implies that

$$\mathscr{H}^g(E) \leq \sum_{n=1}^{\infty} \mathscr{H}^g(E_n).$$

(iii)- We show that $\mathscr{H}^g$ is a Borel measure. Let $E, F \subset X$ such that $\mathrm{d}(E, F) > 0$. Consequently for $A \subset E$ and $B \subset F$, we have $\mathrm{d}(A, B) > 0$ and any centered ν-δ-covering of A is disjoint from any centered ν-δ-covering of B provided $\frac{\mathrm{d}(A, B)}{2} > \delta$. Then, we obtain

$$\overline{\mathscr{H}}^g_\delta(A \cup B) = \overline{\mathscr{H}}^g_\delta(A) + \overline{\mathscr{H}}^g_\delta(B).$$

Letting $\delta \downarrow 0$, one gets

$$\overline{\mathcal{H}}^g(A \cup B) = \overline{\mathcal{H}}^g(A) + \overline{\mathcal{H}}^g(B).$$

Taking the supremum over all $A \subset E$ and all $B \subset F$, it results that

$$\mathcal{H}^g(E \cup F) = \mathcal{H}^g(E) + \mathcal{H}^g(F).$$

Then $\mathcal{H}^g$ is a metric measure and consequently $\mathcal{H}^g$ is a Borel measure.

3. See for example [22]. □

6.3 The Borel Regularity of the Relative Centered Multifractal Measures

This section shows that the relative packing measure $\mathcal{P}^g$ is Borel regular, and if μ and ν are doubling measures, then the usual spherical measure $\mathcal{U}^g$ is equivalent to the relative Hausdorff measure $\mathcal{H}^g$, and thus they define the same dimension. Moreover, this section shows that this statement is true even without the doubling condition if $q \geq 1$ and $t > 0$ or if $q \leq 0$ and $t \leq 0$. In particular, $\mathcal{H}^g$ is a Borel regular measure. We begin with the regularity results for the relative multifractal packing measure $\mathcal{P}^g$.

Theorem 6.1 *The measure $\mathcal{P}^g$ is Borel regular.*

Theorem 6.1 is a consequence from the following lemma.

Lemma 6.1 *Let $\overline{E}$ be the closure of $E \subset X$. Then*

$$\overline{\mathcal{P}}^g(E) = \overline{\mathcal{P}}^g(\overline{E}).$$

Proof By the monotony of $\overline{\mathcal{P}}^g$, we need only prove that $\overline{\mathcal{P}}^g(\overline{E}) \leqslant \overline{\mathcal{P}}^g(E)$. Fix ε, $\delta > 0$ and let $\left\{B(x_i, r_i)\right\}_i$ be a centered ν-δ-packing of $\overline{E}$. By using the continuity, we can choose $s_i > 0$ and $y_i \in B(x_i, s_i) \cap E$ such that

$$\mu\left(B\left(x_i, r_i\right)\right)^q \nu\left(B\left(x_i, r_i\right)\right)^t - \frac{\varepsilon}{2^i} \leqslant \mu\left(B\left(y_i, r_i - \frac{s_i}{2}\right)\right)^q \nu\left(B\left(y_i, r_i - \frac{s_i}{2}\right)\right)^t.$$

In other words

$$g(x_i, r_i) - \frac{\varepsilon}{2^i} \leqslant g\left(y_i, r_i - \frac{s_i}{2}\right). \tag{6.1}$$

Since $B\left(y_i, r_i - \frac{s_i}{2}\right) \subseteq B\left(x_i, r_i\right)$, then $\left(B\left(y_i, r_i - \frac{s_i}{2}\right)\right)_i$ is a centered ν-δ-packing of E, and thus (8.3) gives

$$\begin{aligned}\sum_i g(x_i, r_i) &\leq \sum_i g\left(y_i, r_i - \frac{s_i}{2}\right) + \varepsilon \\ &\leq \overline{\mathcal{P}}^g_\delta(E) + \varepsilon\end{aligned}$$

which implies that

$$\overline{\mathcal{P}}^g_\delta(\overline{E}) \leqslant \overline{\mathcal{P}}^g_\delta(E) + \varepsilon.$$

Next, when ε and δ tends to, we obtain the desired result. □

Proof of Theorem 6.1. It follows from Lemma 6.1 that

$$\mathscr{P}^g(E) = \inf_{E \subset \cup E_i} \sum_i \overline{\mathscr{P}}^g\left(\overline{E}_i\right).$$

Therefore, for each positive integer n, we can choose sets $(E_{ni})_i$ such that $E \subseteq \bigcup_i E_{ni}$ and

$$\sum_i \overline{\mathscr{P}}^g\left(\overline{E}_{ni}\right) \leq \mathscr{P}^g(E) + \frac{1}{n}.$$

Consider the set

$$B = \bigcap_n \bigcup_i \overline{E}_{ni}.$$

It is easy to see that B is Borel and $E \subseteq B$. Finally, we clearly deduce that

$$\mathscr{P}^g(E) \leq \mathscr{P}^g(B) \leq \mathscr{P}^g\left(\bigcup_i \overline{E}_{ni}\right) \leq \sum_i \mathscr{P}^g\left(\overline{E}_{ni}\right) \leq \mathscr{P}^g(E) + \frac{1}{n}, \quad \text{for all } n.$$

Which achieves the proof of Theorem 6.1.

Definition 6.4 [23]Let X be a separable metric space. We say that a measure μ on X fulfils the doubling condition if there are $h, c > 0$ such that for all $x \in \operatorname{spt}\mu$ and $0 < r < h$, we have

$$\mu(B(x, 2r)) \leqslant c\mu(B(x, r)).$$

The following result shows that $\mathscr{H}^g$ is a Borel regular measure even without the doubling condition if $q \geq 1$ and $t > 0$ or if $q \leq 0$ and $t \leq 0$.

Theorem 6.2 *If μ and ν satisfy the doubling condition, then the measure $\mathscr{H}^g$ is Borel regular. Moreover, if $q \leq 0$ and $t \leq 0$ or if $q \geq 1$ and $t > 0$, then $\mathscr{H}^g$ is Borel regular even without the doubling condition.*

We present the tools, as well as the intermediate results, which will be used in the proof of Theorem 6.2.

Lemma 6.2 *If $q \geq 1$ and $t > 0$, then*

$$\mathscr{U}^g(E) = \mathscr{H}^g(E) = \mathscr{P}^g(E) = 0$$

for all $E \subset X$.

Proof Let ξ be the constant that appears in the Besicovitch covering theorem and fix $h > 0$. Let $E \subset B(0, h)$ and $0 < \delta < h$. We consider the following set

$$\mathscr{C} = \left\{B(x, r) \mid x \in E \text{ and } 0 < \nu(B(x, r)) < \delta\right\}.$$

We can find a centered ν-δ-covering $\left\{B\left(x_i, r_i\right)\right\} \subset \mathscr{C}$ of E with the overlap controlled by ξ. Then, one has

$$\begin{aligned}\overline{\mathscr{H}}^g_\delta(E) &\leqslant \sum_i g(x_i,r_i)\\ &\leq \delta^t\left(\sum_i \mu(B(x_i,r_i))\right)^q \quad [\text{ since } q \geqslant 1 \text{ and } t>0]\\ &\leqslant \delta^t \xi^q\left(\mu(B(0,2h))\right)^q.\end{aligned}$$

Letting $\delta \downarrow 0$ shows that $\overline{\mathscr{H}}^g(E) = 0$ for all $E \subset B(0,h)$ which implies that $\mathscr{H}^g(B(0,h)) = 0$ for all $h > 0$. Thus $\mathscr{H}^g(X) = 0$. As every centered covering of a set is also a covering of this set we get the same result for the measure $\mathscr{U}^g$.

Next, for each centered ν-δ-packing $(B(x_i,r_i))_i$ of X we have

$$\begin{aligned}\sum_i g(x_i,r_i) &\leq \delta^t \sum_i \mu(B(x_i,r_i))^q \quad [\text{ since } t>0]\\ &\leq \delta^t\left(\sum_i \mu(B(x_i,r_i))\right)^q \quad [\text{ since } q \geqslant 1]\\ &= \delta^t \mu\left(\bigcup_i B(x_i,r_i)\right)^q\\ &\leq \delta^t.\end{aligned}$$

Which clearly implies that

$$\overline{\mathscr{P}}^g_\delta(X) \leqslant \delta^t.$$

Letting $\delta \searrow 0$ gives $\overline{\mathscr{P}}^g(X) = 0$, and so $\mathscr{P}^g(X) = 0$.

Theorem 6.3 *If μ,ν satisfy the doubling condition, then there exists a constant η depending only on q,t and n such that*

$$\mathscr{U}^g(E) \leqslant \overline{\mathscr{H}}^g(E) \quad \text{and} \quad \mathscr{H}^g(E) \leqslant \eta\,\mathscr{U}^g(E)$$

for all $E \subset X$. In addition, if $q \geq 1$ and $t > 0$ or if $q \leq 0$ and $t \leq 0$ the above inequalities hold even without the doubling condition.

Proof Observe that every centered covering of E is also a covering of E, then the first inequality is obvious. We will need to prove the second inequality.

(i)- Let μ and ν be two measures satisfy the doubling condition and $q > 0$.
Fix $0 < \delta < h$ ($h > 0$ as in Definition 6.4) and let $\left\{B(x_i,r_i)\right\}$ be a ν-δ-covering of E. We can find arbitrarily a ball $B(x,r)$ in $\left\{B(x_i,r_i)\right\}$. Suppose that $B(x,r) \cap E \neq \emptyset$ and let $S := \operatorname{supp}\mu \cap \operatorname{supp}\nu$. If $B(x,r) \cap E \cap \mathrm{S} = \emptyset$ and since X is separable, then we can find a centered ν-δ-covering $\{B(y_k,s_k)\}$ of $E \cap B(x,r)$ with $0 < s_k < d(y_k,\mathrm{S})$. It follows that

$$0 = \sum_k g(y_k,s_k) \leqslant g(x,r).$$

If $B(x,r) \cap E \cap \mathrm{S} \neq \emptyset$, then let $y \in B(x,r) \cap E \cap \mathrm{S}$.

- Case 1 : For $t > 0$, it follows from Definition 6.4 that, there exist two positive constants C and C_1 such that

$$\mu(B(y,2r)) \leqslant C\mu(B(x,r)) \quad \text{and} \quad \nu(B(y,2r)) \leqslant C_1\nu(B(x,r)).$$

Which means to

$$\mu(B(y,2r))^q \leqslant C^q\mu(B(x,r))^q \quad \text{and} \quad \nu(B(y,2r))^t \leqslant C_1^t\nu(B(x,r))^t$$

which implies that

$$g(y,2r) \leqslant C^q C_1^t g(x,r).$$

- Case 2 : For $t \leq 0$, We observe that

$$\nu(B(y,2r))^t \leqslant \nu(B(x,r))^t$$

and it follows from Definition 6.4 that, there exists a positive constant C such that

$$\mu(B(y,2r)) \leqslant C\mu(B(x,r))$$

and thus

$$\mu(B(y,2r))^q \leqslant C^q\mu(B(x,r))^q.$$

Which implies that

$$g(y,2r) \leqslant C^q g(x,r).$$

Here there exist $\eta > 0$ and $\beta \geq 2$ such that $\left\{B(y_i,s_i)\right\}$ is a centered ν-$\beta\delta$-covering of E and

$$\sum_i g(y_i,s_i) \leqslant \eta \sum_i g(x_i,r_i).$$

Then we can conclude that

$$\overline{\mathscr{H}}^g_{\beta\delta}(E) \leqslant \eta\mathscr{U}^g_\delta(E).$$

Take $\delta \downarrow 0$ in the above expression, we get the result.

(ii)- For $q \leqslant 0$ and $t \leqslant 0$. Observe that

$$\mu(B(x,2r))^q \leqslant \mu(B(x,r))^q \quad \text{and} \quad \nu(B(x,2r))^t \leqslant \nu(B(x,r))^t,$$

which implies that the above proof works even without the doubling condition.

(iii)- For $q \leqslant 0$ and $t > 0$. The proof is identical to the proof of (i)-case 1 and is therefore omitted. □

Lemma 6.3 *Let $E \subset X$ and G be a Borel set such that $E \subset G$. Then we can find a Borel set $E \subset F \subset G$ such that*

$$\overline{\mathscr{H}}^g(E) = \overline{\mathscr{H}}^g(F).$$

Proof For each $n \in \mathbb{N}$, we can choose a centered ν-$\frac{1}{n}$-covering $\left\{B(x_{ni},r_{ni})\right\}_i$ of E such that

$$\sum_i g(x_{ni},r_{ni}) \leqslant \overline{\mathscr{H}}^g_{\frac{1}{n}}(E) + \frac{1}{n} \quad \text{and} \quad B(x_{ni},r_{ni}) \cap E \neq \varnothing \quad \text{for all} \quad n,i \in \mathbb{N}.$$

Now, we consider the following set

$$F = \bigcap_n \bigcup_i B\left(x_{ni}, r_{ni}\right) \cap G.$$

Therefore F is a Borel set and $E \subset F \subset G$.

We will now prove that

$$\overline{\mathscr{H}}^g(E) = \overline{\mathscr{H}}^g(F).$$

As all $B\left(z_{ni}, \kappa_{ni}\right)$ are centered in E, they are also centered in F and consequently, we get $\overline{\mathscr{H}}^g(F) \leq \overline{\mathscr{H}}^g(E)$. In order to prove the other inequality, let $\left\{B\left(x_i, \kappa_i\right)\right\}_i$ be a centered ν-δ-covering of F. Let $\varepsilon > 0$ and $B(x, \kappa) \in \left\{B\left(x_i, \kappa_i\right)\right\}_i$. It follows from $\mu\left(B(x, \kappa)\right) < \infty$ and $\nu\left(B(x, \kappa)\right) < \infty$ that, we can find $\kappa < \kappa_1 < 2\kappa$ and $\kappa < \kappa_2 < 2\kappa$ such that

$$\mu\left(B(x, \kappa_1)\right)^q \leqslant \mu\left(B(x, \kappa)\right)^q + \frac{\varepsilon}{2^i} \quad \text{and} \quad \nu\left(B(x, \kappa_2)\right)^t \leqslant \nu\left(B(x, \kappa)\right)^t + \frac{\varepsilon}{2^i}. \tag{6.2}$$

Since E is dense in F, we can find $\epsilon, \varepsilon_1, \varepsilon_2, \varepsilon_3 > 0$, $y \in E$ and $\kappa < r < \kappa_0 := \min(\kappa_1, \kappa_2)$ with

$$B(x, \kappa) \subset B(y, r) \subset B\left(x, \kappa_0\right).$$

(i)- For $q \leq 0$ and $t \leq 0$, one has

$$\mu\left(B(y, r)\right)^q \leqslant \mu\left(B(x, \kappa)\right)^q \quad \text{and} \quad \nu\left(B(y, r)\right)^t \leqslant \nu\left(B(x, \kappa)\right)^t.$$

It follows that

$$\mu\left(B(y, r)\right)^q \nu\left(B(y, r)\right)^t \leqslant \mu\left(B(x, \kappa)\right)^q \nu\left(B(x, \kappa)\right)^t.$$

(ii)- For $q > 0$ and $t > 0$, by using (6.2) we get

$$\begin{aligned} \mu\left(B(y, r)\right)^q &\leq \mu\left(B(x, \kappa_1)\right)^q \\ &\leq \mu\left(B(x, \kappa)\right)^q + \frac{\varepsilon}{2^i} \end{aligned}$$

and

$$\begin{aligned} \nu\left(B(y, r)\right)^t &\leq \mu\left(B(x, \kappa_2)\right)^t \\ &\leq \nu\left(B(x, \kappa)\right)^t + \frac{\varepsilon}{2^i}. \end{aligned}$$

Which means that

$$\mu\left(B(y, r)\right)^q \nu\left(B(y, r)\right)^t \leqslant \mu\left(B(x, \kappa)\right)^q \nu\left(B(x, \kappa)\right)^t + \frac{\varepsilon_1}{2^i}.$$

(iii)- For $q > 0$ and $t \leq 0$, it follows from (6.2) that

$$\begin{aligned} \mu\left(B(y, r)\right)^q &\leq \mu\left(B(x, \kappa_1)\right)^q \\ &\leq \mu\left(B(x, \kappa)\right)^q + \frac{\varepsilon}{2^i} \end{aligned}$$

and

$$\nu\left(B(y,r)\right)^t \leqslant \nu\left(B(x,\kappa)\right)^t.$$

Therefore

$$\mu\left(B(y,r)\right)^q \nu\left(B(y,r)\right)^t \leqslant \mu\left(B(x,\kappa)\right)^q \nu\left(B(x,\kappa)\right)^t + \frac{\varepsilon_2}{2^i}.$$

(iv)- For $q \leq 0$ and $t > 0$. Using similar arguments for (iii), we have

$$\mu\left(B(y,r)\right)^q \nu\left(B(y,r)\right)^t \leqslant \mu\left(B(x,\kappa)\right)^q \nu\left(B(x,\kappa)\right)^t + \frac{\varepsilon_3}{2^i}.$$

It both cases, we have

$$\mu\left(B(y,r)\right)^q \nu\left(B(y,r)\right)^t \leqslant \mu\left(B(x,\kappa)\right)^q \nu\left(B(x,\kappa)\right)^t + \frac{\epsilon}{2^i}.$$

Applying the above to our centered covering, then there exists $\beta \geq 2$ such that $\{B(y_i, r_i)\}$ is a centered ν-$\beta\delta$-covering of E with

$$\sum_i g(y_i, r_i) \leqslant \sum_i g(x_i, s_i) + \epsilon.$$

This yields

$$\overline{\mathscr{H}}^g_{\beta\delta}(E) \leqslant \overline{\mathscr{H}}^g_{\delta}(F) + \epsilon$$

and when δ and ϵ tends to 0 in the above expression, we get the result. □

Lemma 6.4 *Assume that there exists a constant η such that for every $E \subset X$,*

$$\mathscr{U}^g(E) \leqslant \overline{\mathscr{H}}^g(E) \quad \text{and} \quad \mathscr{H}^g(E) \leqslant \eta\,\mathscr{U}^g(E).$$

Then $\mathscr{H}^g$ is Borel regular.

Proof Let $E \subset X$ and we may assume that $\mathscr{H}^g(E) < \infty$ which implies that $\mathscr{U}^g(E) < \infty$. It follows from Proposition 6.1-(3) that, we can find a Borel set $F \supset E$ such that $\mathscr{U}^g(F) = \mathscr{U}^g(E)$. We will prove that the set F is the required Borel set. Then we suppose, on the contrary, that $\mathscr{H}^g(F) > \mathscr{H}^g(E)$. It follows from the definition that, there exists $C \subset F$ such that $\overline{\mathscr{H}}^g(C) > \mathscr{H}^g(E)$. By using Lemma 6.3, we may assume that C is Borel. We clearly deduce that

$$\overline{\mathscr{H}}^g(C) > \overline{\mathscr{H}}^g(C \cap E). \tag{6.3}$$

Lemma 6.3 implies that there exists a Borel set $C \cap E \subset D \subset C$ such that

$$\overline{\mathscr{H}}^g(C \cap E) = \overline{\mathscr{H}}^g(D). \tag{6.4}$$

Since $C \backslash D \subset F \backslash E$ and $C \backslash D$ is Borel, we have

$$\mathscr{U}^g(F) = \mathscr{U}^g(F\backslash(C\backslash D)) + \mathscr{U}^g(C\backslash D) \geqslant \mathscr{U}^g(E) + \mathscr{U}^g(C\backslash D) \geqslant \mathscr{U}^g(E) = \mathscr{U}^g(F).$$

Therefore $\mathscr{U}^g(C\backslash D) = 0$ which gives that $\overline{\mathscr{H}}^g(C\backslash D) = 0$. It follows from (6.3), (6.4) and Proposition 6.1-(1) that

$$\overline{\mathscr{H}}^g(C) \leqslant \overline{\mathscr{H}}^g(D) + \overline{\mathscr{H}}^g(C\backslash D) = \overline{\mathscr{H}}^g(D) = \overline{\mathscr{H}}^g(C \cap E) < \overline{\mathscr{H}}^g(C)$$

which a contradiction. □

Proof of Theorem 6.2. Follows immediately from Theorem 7.2 and Lemma 6.4.

6.4 Remarks and Discussion

The result in Theorem 6.1 and the first assertion in Theorem 6.2 hold if we replace the centered ν-δ-coverings (ν-δ-packings) by the centered δ-coverings (δ-packings), for more details see [21]. By using a similar technique of [27] we will construct two measures μ and ν in order to show that Theorem 6.2 does not necessarily hold without the assumptions stated there. More precisely, let μ be the measure for constructed in [27] and ν is basically the Lebesgue measure, then by setting $g(x,r) = \mu(B(x,r))^{\frac{1}{2}}\nu(B(x,r))^0$, we obtain that the Borel regularity of the centered relative Hausdorff measure cannot work for non-doubling measures, $q = \frac{1}{2}$ and $t = 0$. Though the same type of construction can be done for all $0 < q < 1$. More precisely, if we assume that ν is basically the Lebesgue measure, then the above results reduce to Schechter's theorems in [27].

We don't need the assumption that the μ and ν are finite on balls but without some similar conditions, the measures and they could be infinite on all nonempty sets. This would cause unnecessary technical difficulties later, so we excluded this extreme case in the above definitions. Considering remarks that the definition of these measures depends very much on the metric on the underlying set, even though for clarity reasons we have decided not to add this extra parameter in the symbols. However, as two different metrics on the same set usually define two different Hausdorff (spherical and centered Hausdorff) measures, the reader should keep this fact in mind throughout this chapter. The reader should also keep in mind that we have simply written g for gauge functions $g^{q,t}_{\mu,\nu}$ depending on μ, ν, q and t. We believe that putting too many different variables in formulas is rather unpleasant for the reader and that omitting these extra parameters will create no confusion. Also, note that the set function $\overline{\mathscr{H}}^g$ is not necessarily monotone, since the smaller set may not have the center points for the ideal covering.

The equivalence between the usual spherical measure and the classical centered Hausdorff measure is being used to show the Borel regularity of the centered Hausdorff measure for the cases described before. This is an essential basic property of measures, which was unknown for this measure for many years. The interesting question of whether this regularity property is at all true for the centered Hausdorff measure has been asked in various works as [18, 19]. The theory that is being developed to give an affirmative answer to this problem is based on [27].

6.5 Concluding Remarks

The present work deals with a Borel regularity of the centered relative multifractal measures $\mathscr{P}^g$ and $\mathscr{H}^g$. It is proved that $\mathscr{P}^g$ is Borel regular. It is also shown that the usual spherical measure $\mathscr{U}^g$ and the relative Hausdorff measure $\mathscr{H}^g$ are equivalent in the case where μ and ν satisfy the doubling condition. In addition, it is shown that this statement is true even without the doubling condition if $q \geq 1$ and $t > 0$ or if $q \leq 0$ and $t \leq 0$ which implies, in particular, $\mathscr{H}^g$ is Borel regular measure. Finally, a counter-example shows that the proof for the Borel regularity of the measure $\mathscr{H}^g$ cannot work for non-doubling measures and $0 < q < 1$.

Acknowledgments This work was supported by Analysis, Probability & Fractals Laboratory (No: LR18ES17).

References

[1] Attia, N. and B. Selmi. 2019. Relative multifractal box-dimensions. Filomat, 33: 2841–2859.

[2] Attia, N., B. Selmi and Ch. Souissi. 2017. Some density results of relative multifractal analysis. Chaos, Solitons & Fractals, 103: 1–11.

[3] Ben Mabrouk, A. and A. Farhat. 2021. A mixed multifractal analysis for quasi Ahlfors vector-valued measures. Fractals Accepted, ArXiv:2103.05466v1.

[4] Ben Mabrouk, A. and A. Farhat. 2021. Mixed multifractal densities for quasi-Ahlfors vector-valued measures, Fractals Accepted, ArXiv:2103.05246v1.

[5] Billingsley, P. 1960. Hausdorff dimension in probability theory[1], Illinois. J. Math. 4: 187–209.

[6] Billingsley, P. 1961. Hausdorff dimension in probability theory II, Illinois. J. Math. 5: 291–298.

[7] Carathéodory, C. 1914. Uber das lineare Maß von Punktmengen, eine Verallgemeinerung des Längenbegriffs. Nach. Ces. Wiss. Gottingen, 1914: 406–426.

[8] Cole, J. 2000. Relative multifractal analysis. Choas, Solitons & Fractals, 11: 2233–2250.

[9] Dai, C. 1995. Frostman lemma for Hausdorff measure and Packing measure in probability spaces. J Nanjing University, Mathematical Biquarterly, 12: 191–203.

[10] Dai, C. and Y. Hou. 2005. Frostman lemmas for Hausdorff measure and packing measure in a product probability space and their physical application. Chaos, Solitons and Fractals, 24: 73–744.

[11] Dai, C. and Y. Li. 2006. A multifractal formalism in a probability space. Chaos, Solitons and Fractals, 27: 5–73.

[12] Dai, C. and Y. Li. 2007. Multifractal dimension inequalities in a probability space. Chaos, Solitons and Fractals, 34: 213–223.

[13] Dai, M., X. Peng and W. Li. 2010. Relative multifractal analysis in a probability space. Internal Journal of Nonlinear Science, 10: 313–319.

[14] Das, M. 2008. Billingsley's packing dimension. Proceedings of the American Mathematical Society, 136: 273–278.

[15] Das, M. 2005. Hausdorff measures, dimensions and mutual singularity. Transactions of the American Mathematical Society, 357: 4249–4268.

[16] Douzi, Z. and B. Selmi. 2016. Multifractal variation for projections of measures. Chaos, Solitons and Fractals, 91: 414–420.

[17] Douzi, Z. and B. Selmi. 2021. A relative multifractal analysis: Box-dimensions, densities, and projections. Quaestiones Mathematicae. doi.org/10.2989/16073606.2021.1941375.

[18] Edgar, G.A. 1995. Fine variation and fractal measures. Real Anal. Exchange, 20: 256–280.

[19] Edgar, G.A. 1998. Integral, Probability, and Fractal Measure Springer-Verlag, New York.

[20] Hausdorff, F. 1919. Dimension und Äußeres Maß. Math. Ann., 79: 157–179.

[21] Khelifi, M., H. Lotfi, A. Samti and B. Selmi. 2020. A joint multifractal analysis. Choas, Solitons & Fractals, 140: 110091.

[22] Mattila, P. 1995. Geometry of Sets and Measures in Euclidian Spaces: Fractals and Rectifiability. Cambridge University Press.

[23] Olsen, L. 1995. A multifractal formalism. Advances in Mathematics, 116: 82–196.

[24] Peyrière, J. 2004. A vectorial multifractal formalism. Proc. Sympos, Pure Math., 72: 217–230.

[25] Raymond, X.S. and C. Tricot. 1988. Packing regularity of sets in n-space. Math. Proc. Camb. Philos. Soc., 103: 133–145.

[26] Rogers, C.A. 1998. Hausdorff Measures, Reprint of the 1970 original. Cambridge University Press, Cambridge.

[27] Schechter, A. 2000. On the centred Hausdorff measure. Journal of the London Mathematical Society, 62: 843–851.

[28] Selmi, B. 2019. On the strong regularity with the multifractal measures in a probability space. Anal. Math. Phys., 9: 1525–1534.

[29] Selmi, B. 2018. Measure of relative multifractal exact dimensions. Adv. Appl. Math. Sci., 17: 629–643.

[30] Selmi, B. 2020. The relative multifractal analysis, review and examples. Acta Scientiarum Mathematicarum, 86: 635–666.

[31] Selmi, B. 2021. The relative multifractal densities: A review and application. Journal of Interdisciplinary Mathematics, 24(6): 1627–1644.

[32] Tricot, C. 1982. Two definitions of fractional dimension. Math. Proc. Cambridge Philos. Soc., 91: 54–74.

Chapter 7

A Mixed Multifractal Analysis of Vector-valued Measures: Review and Extension to Densities and Regularities of Non-necessary Gibbs Cases

Anouar Ben Mabrouk[1] and Bilel Selmi[2,*]

7.1 Introduction

In the multifractal analysis of measures, the study of the behavior of the measure is usually transformed into a study of sets related to the local behavior of such measures called level sets and defined according to the so-called Holder regularity of the measure. The focuses thus may somehow forget about the measure and its point-wise character and falls in set theory and the suitable coverings that permit the computation of the Hausdorff dimension. However, some geometric sets are essentially utilised by measures that are supported by them, i.e., given a set E and a measure μ, the quantity $\mu(E)$ may be computed as the maximum value $\mu(F)$ for all subsets $F \subset E$. So, contrary to the previous idea, we mathematically forget the geometric structure of E and focus instead on the properties of the measure μ. The set E is thus partitioned into α-level sets $X_\mu(\alpha)$ relatively to the regularity exponent of μ.

This makes the including of the measure μ into the computation of the Hausdorff (or fractal) dimension and thus into the definition of the Hausdorff measure a necessity to understand more the geometry of the set simultaneously with the behavior of the measure that it is supported on. One step ahead in this direction has been conducted by Olsen in [87] where the author introduced multifractal generalizations of the fractal dimensions such as Hausdorff, packing and Bouligand ones by considering general variants of measures.

A first step in the mixed multifractal analysis has been developed by the same author for one very restrictive class of measures known as the self-affine measures [90] dealing precisely with Rényi dimensions for finitely many self-affine measures. Next, motivated by this study, a mixed multifractal analysis has been developed in [14] for vector-valued measures in some more general contexts. By assuming a restrictive hypothesis looking like Gibbs-type measures and by proving a general mixed large deviation formalism a mixed multifractal formalism has been proved. In [14] and [15] a mixed multifractal analysis inspired from the one for measures has been developed in the functional case. By concentring a vector-valued Gibbs-like measure on the singularities set of finitely and simultaneously many functions, a mixed multifractal formalism for functions has been developed. General results for almost all functions have been proved and a mixed multifractal formalism has been proved for self-similar quasi self-similar functions as well as their superpositions (which are not self-similar neither quasi self-similar). For more details and backgrounds on multifractal analysis as well as the mixed generalizations the readers may be referred also to the following essential references [14], [90], [115], [117], [125], [127], [128], [129], [130], [132], [133].

[1] Algebra, Number Theory and Nonlinear Analysis Laboratory LR18ES15, Department of Mathematics, Faculty of Sciences, University of Monastir, 5000 Monastir, Tunisia.

[2] Analysis, Probability & Fractals Laboratory LR18ES17, Department of Mathematics, Faculty of Sciences of Monastir, University of Monastir, 5000-Monastir, Tunisia.
Email: anouar.benmabrouk@fsm.rnu.tn

[*] Corresponding author: bilel.selmi@fsm.rnu.tn,bilel.selmi@isetgb.rnu.tn

The concept of mixed multifractal analysis had been firstly applied twenty years ago in the framework of turbulence without focusing on the fundamental mathematical formulation [83]. The idea that has been issued from the mixed models in statistics and data analysis, where mixed analysis is already known and applied in some simple cotexts. We may recall here that the co-variance of statistical series is in fact a type of mixed measure.

Next, with the appearance and the rapid developments of fractal/multifractal analysis, and its merging in quasi all branches of science, especially those related to financial time series, the need for a mixed form of the original multifractal analysis has become a necessity.

A first indirect step has been conducted by Olsen in [87] by introducing multifractal generalizations of the Hausdorff and packing measures relative to Borel probability measures on the Euclidean space, consisting of correlating the powers of the Lebesgue measure (diameter of the balls, or the covering elements) with an extra Borel probability measure, and vice-versa. Let μ be a Borel probability measure on $\mathbb{R}^d$, Olsen considered the multifractal measures based on the summations

$$\sum_i \mu(B(x_i, r_i))^q r_i^t,$$

where $(B(x_i, r_i))_i$ are centered coverings or centered packings according to the measure to be studied and applied, $(q, t) \in \mathbb{R}^2$. This means that the correlation function

$$h_{q,t}(r) = \mu(B(x, r))^q r^t, \quad r > 0,$$

has been applied. By assuming that the measure μ lies in the class of doubling, self-similar, and Gibbs measures, the next step consisted in constructing a Gibbs measure on the α-singularities sets

$$X_\mu(\alpha) = \left\{ x \middle| \lim_{r \downarrow 0} \frac{\log(\mu(B(x, r)))}{\log r} = \alpha \right\}$$

for a suitable choice of α sufficient to cover all the support of the measure. The main tool was the application of the large deviation formalism and Billingsley theory.

A step forward already in the path of mixed multifractal analysis of measures has been developed in [16] by involving an extra gauge function in the correlating set function, by considering summations of the form

$$\sum_i \mu(B(x_i, r_i))^q r_i^t e^{\varphi(r)},$$

where the function φ behaves as $o(\log)$ near zero. This is in fact a weak point as it means that the measure μ is compared again to the powers of the diameter, and the inclusion of the extra function φ is only needed for some flexibility in the estimations around the dimension associated. Moreover, there is no variable power on this extra term.

An early investigation has also applied the correlation functions, and thus used a mixed variant of multifractal measure indirectly, dealing precisely with the so-called Carathéodory measure and dimension. In [101], the concept of dimension is introduced with a special structure known as the C-structure. Let $\mathcal{F}$ be a collection of subsets of a space X and ξ, η and ψ be defined on $\mathcal{F}$ and positive. Under suitable assumptions on these functions, the (outer) Carathéodory measure has been defined as the summation

$$\mathcal{H}_\alpha(X) = \inf \left\{ \sum_{U \in \mathcal{G}} \xi(U) \eta(U)^\alpha \right\}.$$

For special choices of the functions ξ, η and ψ, the space X and the collection $\mathcal{F}$, this measure permits us to obtain the well-known Hausdorff and packing measures (see also [100]).

The new framework to be reviewed here consists of replacing all the Lebesgue type measures (diameters, log) in one merged form, by involving a control gauge function, that permits control of many measures at the same time. This is possible by considering summations of the form

$$\sum_i \mu_1(B(x_i,r_i))^{q_1} \dots \mu_k(B(x_i,r_i))^{q_k} e^{t\varphi(r_i)},$$

for $q_1,\dots,q_k,t \in \mathbb{R}$ and $\mu_1,\dots,\mu_k$ Borel probability measures on $\mathbb{R}^d$ and $\varphi : (0,1) \rightarrow (-\infty,0)$ is some suitable function. Associated Hausdorff and packing measures are introduced in [82] with eventual dimensions. Considering the remark that the original casae due to [87] is simply obtained by setting $k = 1$ and $\varphi(r) = \log r$. It can be remarked also that the estimation of the multifractal spectrum in the present extension is based on the function φ tailored to yield more general multifractal decomposition sets

$$X_{\mu,\varphi}(\alpha) = \left\{ x \middle| \lim_{r\downarrow 0} \frac{\log(\mu(B(x,r)))}{\varphi(r)} = \alpha \right\}.$$

Another motivation comes from the discrete fractal theory that has been introduced recently in mathematics and some applied fields. We know very well that the usual fractal analysis did not give important information in the context of discrete spaces, as these are zero dimension sets relative to the Hausdorff and packing dimensions in their original definitions. This makes it necessary to introduce appropriately a discrete multifractal analysis. These measures cannot be compared to power-laws diameters. Many applications have been discussed for the utility of discrete multifractal analysis in many fields such as number theory, random walks, etc. (see [4], [7], [60]).

In this chapter, we are concerned with the multifractal analysis of measures in a mixed case (which can already be adapted to single cases) where the hypothesis of the existence of Gibbs-like and/or doubling measures which is supported by the singularities sets is relaxed. We aim to consider some cases of simultaneous behaviors of measures where the local Hölder behavior is controlled by a special and suitable function that allows the extra-hypothesis of Gibbs-like measures not to be necessary.

In the present chapter, we are precisely focusing on some density estimations of vector-valued measures in a general framework of mixed multifractal analysis for measures that have been developed recently in pure mathematics to extend the multifractal analysis. However, in the applied point of view, mixed multifractal analysis has been developed independently especially in physics and statistics ([52], [83]). We propose to extend the existing cases of mixed multifractal analysis of measures to the more general context that will englobe all the existing cases as special ones by introducing a mixed density notion where the hypothesis of being Gibbs for the measures is no longer assumed. For background, we may refer to [2], [3], [32], [33], [40], [81], [82], [83].

The problem studied here has been the subject of several studies. In [3], some multifractal densities in the framework of single multifractal analysis have been investigated. Motivated by [3], an extending study has been developed in [49] where the hypotheses on the applied measure have been revised. The theory was next applied to a case of quasi Ahlfors regular vector-valued measures introduced in [81]. Next, in [82] a more general mixed (and thus single) multifractal analysis for vector-valued measures has been developed extending all the existing cases ([33], [48], [87], [128]) and where the hypothesis of being Gibbs-like is no longer necessary for the applied measures. Instead, a gauge control function φ is included in the definition of the Hausdorff and packing measures to best control the local behavior of the vector-valued measure μ. In the present paper, we aim to consider the last context of mixed multifractal analysis developed in [82] and to develop some mixed multifractal densities estimations extending the results of [3], [18], [33], [41], [42], [49], [78], [87], [88], [90], [102], [103], [108], [109], [110].

The next section is devoted to the presentation of the general settings of the present study. Section 3 is devoted to the main results. In Section 4, results on the regularities of the φ-mixed multifractal generalizations of Hausdorff and packing measures are developed leading to the decomposition theorem of Besicovitch's type. Section 5 is concerned with an application of the previous results in which a necessary condition for a strong regularity with the φ-mixed multifractal generalizations of Hausdorff and packing measures has been established. Sections 6 and 7 contain some remarks, a discussion, and a conclusion. Section 8 is an appendix in which some useful well-known theorems have been recalled.

7.2 The φ-mixed Multifractal Measures and Dimensions

In this section, we review briefly the φ-mixed multifractal generalizations of the Hausdorff and packing measures and dimensions already developed in [82] and [112], and thus we introduce the general settings and context for our study to be developed next.

Let $k \in \mathbb{N}$ be a fixed integer and $\mu = (\mu_1, \mu_2, \ldots, \mu_k)$ be a vector-valued measure composed of Borel probability measures on $\mathbb{R}^d$. Let also $\varphi : \mathbb{R}_+ \to \mathbb{R}$ be such that

$$\varphi \text{ is non-decreasing and } \lim_{r \downarrow 0} \varphi(r) = -\infty. \tag{7.1}$$

For $x \in \mathbb{R}^d$ and $r > 0$ we denote $B(x, r)$ the ball of radius r and center x and

$$\mu(B(x,r)) \equiv \big(\mu_1(B(x,r)), \ldots, \mu_k(B(x,r))\big)$$

and for $\mathbf{q} = (q_1, q_2, \ldots, q_k) \in \mathbb{R}^k$, we write

$$(\mu(B(x,r)))^{\mathbf{q}} \equiv (\mu_1(B(x,r)))^{q_1} \ldots (\mu_k(B(x,r)))^{q_k}.$$

For $E \subseteq \mathbb{R}^d$ nonempty, $\epsilon > 0$ and $t \in \mathbb{R}$ consider the quantity

$$\overline{\mathcal{H}}^{\mathbf{q},t}_{\mu,\varphi,\epsilon}(E) = \inf\left\{ \sum_i (\mu(B(x_i, r_i)))^{\mathbf{q}} e^{t\varphi(r_i)} \right\},$$

where the inf is taken over the set of all centered ϵ-coverings of E, and for the empty set, $\overline{\mathcal{H}}^{\mathbf{q},t}_{\mu,\epsilon}(\emptyset) = 0$. Consider next

$$\overline{\mathcal{H}}^{\mathbf{q},t}_{\mu,\varphi}(E) = \lim_{\epsilon \downarrow 0} \overline{\mathcal{H}}^{\mathbf{q},t}_{\mu,\varphi,\epsilon}(E).$$

The φ-mixed multifractal generalization of the Hausdorff measure is

$$\mathcal{H}^{\mathbf{q},t}_{\mu,\varphi}(E) = \sup_{F \subseteq E} \overline{\mathcal{H}}^{\mathbf{q},t}_{\mu,\varphi}(F).$$

Similarly, we introduce the φ-mixed multifractal generalization of the packing measure as follows. Let

$$\overline{\mathcal{P}}^{\mathbf{q},t}_{\mu,\varphi,\epsilon}(E) = \sup\left\{\sum_i (\mu(B(x_i,r_i)))^{\mathbf{q}} e^{t\varphi(r_i)}\right\}$$

where the sup is taken over the set of all centered ϵ-packings of E. For the empty set, we set as usual $\overline{\mathcal{P}}^{\mathbf{q},t}_{\mu,\varphi,\epsilon}(\emptyset) = 0$. Next, we denote

$$\overline{\mathcal{P}}^{\mathbf{q},t}_{\mu,\varphi}(E) = \lim_{\epsilon\downarrow 0} \overline{\mathcal{P}}^{\mathbf{q},t}_{\mu,\varphi,\epsilon}(E) = \inf_{\epsilon>0} \overline{\mathcal{P}}^{\mathbf{q},t}_{\mu,\varphi,\epsilon}(E)$$

and finally we obtain the φ-mixed multifractal generalization of the packing measure as

$$\mathcal{P}^{\mathbf{q},t}_{\mu,\varphi}(E) = \inf_{E\subseteq\cup_i E_i} \sum_i \overline{\mathcal{P}}^{\mathbf{q},t}_{\mu,\varphi}(E_i).$$

The following theorem resumes some properties of the φ-mixed multifractal measure $\mathcal{H}^{\mathbf{q},t}_{\mu,\varphi}$ introduced above.

Theorem 7.1 *$\mathcal{H}^{\mathbf{q},t}_{\mu,\varphi}$ and $\mathcal{P}^{\mathbf{q},t}_{\mu,\varphi}$ are outer metric measures on $\mathbb{R}^d$.*

The proof of this result for $\mathcal{H}^{\mathbf{q},t}_{\mu,\varphi}$ is technic and follows standard arguments. However, the proof for $\mathcal{P}^{\mathbf{q},t}_{\mu,\varphi}$ is more specific. To prove the theorem, we recall firstly the following definition.

Definition 7.1 Let X be a non empty set, and denote $\mathcal{P}(X)$ be its power set. A set function $\Phi : \mathcal{P}(X) \rightarrow \mathbb{R}$ is said to be an outer measure on X, if it satisfies the following three assertions,

a) $\Phi(\emptyset) = 0$.
b) Φ is monotone, in the sense that, for any subsets $E \subseteq F \subset X$, we have

$$\Phi(E) \leq \Phi(F).$$

c) Φ is sub-additive, in the sense that, for any sequence $(A_n)_n$ of subsets of X, we have

$$\Phi(\bigcup_n A_n) \leq \sum_n \Phi(A_n).$$

Proof of Theorem 7.1. We will split the proof into four parts.

1. $\mathcal{H}^{\mathbf{q},t}_{\mu,\varphi}$ is an outer measure on $\mathbb{R}^d$.
2. $\mathcal{H}^{\mathbf{q},t}_{\mu,\varphi}$ is metric on $\mathbb{R}^d$.
3. $\mathcal{P}^{\mathbf{q},t}_{\mu,\varphi}$ is an outer measure on $\mathbb{R}^d$.
4. $\mathcal{P}^{\mathbf{q},t}_{\mu,\varphi}$ is metric on $\mathbb{R}^d$.

So, let us develop each point.
3) We shall show the three properties of Definition 7.1 for $\Phi = \mathcal{H}^{\mathbf{q},t}_{\mu,\varphi}$.
1.a) The equality is $\mathcal{H}^{\mathbf{q},t}_{\mu,\varphi}(\emptyset) = 0$ is obvious due to the construction convention assumed for $\overline{\mathcal{H}}^{\mathbf{q},t}_{\mu,\varphi}$, and $\mathcal{H}^{\mathbf{q},t}_{\mu,\varphi}$.
1.b) Let $E \subseteq F$ be two subsets of $\mathbb{R}^d$. It holds that

$$\mathcal{H}^{\mathbf{q},t}_{\mu,\varphi}(E) = \sup_{A\subseteq E} \overline{\mathcal{H}}^{\mathbf{q},t}_{\mu,\varphi}(A) \leq \sup_{A\subseteq F} \overline{\mathcal{H}}^{\mathbf{q},t}_{\mu,\varphi}(A) = \mathcal{H}^{\mathbf{q},t}_{\mu,\varphi}(F).$$

As a result,

$$\mathcal{H}^{\mathbf{q},t}_{\mu,\varphi}(E) \leq \mathcal{H}^{\mathbf{q},t}_{\mu,\varphi}(F).$$

1.c) Let $(E_n)_n$ be a sequence of subsets of $\mathbb{R}^d$ such that

$$\sum_n \mathcal{H}^{\mathbf{q},t}_{\mu,\varphi}(E_n) < \infty.$$

Consider next $\eta, \epsilon > 0$, and $(B(x_{ni}, r_{ni}))_i$ a centred η-covering of E_n, for which, we have

$$\sum_i \mu(B(x_{ni}, r_{ni}))^q e^{t\varphi(r_{ni})} \leq \overline{\mathcal{H}}^{q,t}_{\mu,\varphi,\eta}(E_n) + \frac{\epsilon}{2^n}.$$

It holds easily that $(B(x_{ni}, r_{ni}))_{n,i}$ is a centered η-covering of $\bigcup_n E_n$. Consequently,

$$\begin{aligned}
\overline{\mathcal{H}}^{\mathbf{q},t}_{\mu,\varphi,\eta}\left(\bigcup_n E_n\right) &\leq \sum_n \sum_i (\mu(B(x_{ni}, r_{ni})))^{\mathbf{q}} e^{t\varphi(r_{ni})} \\
&\leq \sum_n \left(\overline{\mathcal{H}}^{\mathbf{q},t}_{\mu,\varphi,\eta}(E_n) + \frac{\epsilon}{2^n}\right) \\
&\leq \sum_n \left(\overline{\mathcal{H}}^{\mathbf{q},t}_{\mu,\varphi}(E_n) + \frac{\epsilon}{2^n}\right) \\
&\leq \sum_n \mathcal{H}^{\mathbf{q},t}_{\mu,\varphi}(E_n) + \epsilon.
\end{aligned}$$

Next, as $\eta, \epsilon \downarrow 0$, we obtain

$$\overline{\mathcal{H}}^{\mathbf{q},t}_{\mu,\varphi}(\bigcup_n E_n) \leq \sum_n \mathcal{H}^{\mathbf{q},t}_{\mu,\varphi}(E_n).$$

Consider next a covering $(A_n)_n$ of F by subsets of $\mathbb{R}^d$. We immediately observe that

$$\begin{aligned}
\overline{\mathcal{H}}^{\mathbf{q},t}_{\mu,\varphi}(F) &= \overline{\mathcal{H}}^{\mathbf{q},t}_{\mu,\varphi}\left(\bigcup_n (A_n F)\right) \\
&\leq \sum_n \mathcal{H}^{\mathbf{q},t}_{\mu,\varphi}(A_n \cap F) \\
&\leq \sum_n \mathcal{H}^{\mathbf{q},t}_{\mu,\varphi}(A_n).
\end{aligned}$$

By taking the upper bound on all sets $F \subseteq \bigcup_n A_n$, we obtain

$$\mathcal{H}^{\mathbf{q},t}_{\mu,\varphi}(\bigcup_n A_n) \leq \sum_n \mathcal{H}^{\mathbf{q},t}_{\mu,\varphi}(A_n).$$

2) We shall show that $\mathcal{H}^{\mathbf{q},t}_{\mu,\varphi}$ is a metric measure on $\mathbb{R}^d$. To do it, let E_1, E_2 be two subsets of $\mathbb{R}^d$, satisfying

$$d(E_1, E_2) > 0, \quad \text{and} \quad \mathcal{H}^{\mathbf{q},t}_{\mu,\varphi}(E_1 \cup E_2) < \infty,$$

where $d(.,.)$ is the Euclidean distance on $\mathbb{R}^d$. Consider next a positive real number ϵ such that $0 < \epsilon < d(E_1, E_2)$. Let finally $\eta > 0$, $F_1 \subseteq E_1$, $F_2 \subseteq E_2$, and $(B(x_i, r_i))_i$ be a centered ϵ-covering of the disjoint union $F_1 \cup F_2$ satisfying

$$\overline{\mathcal{H}}^{q,t}_{\mu,\varphi,\epsilon}(F_1 \cup F_2) \leq \sum_i (\mu(B(x_i, r_i)))^q e^{t\varphi(r_i)} \leq \overline{\mathcal{H}}^{q,t}_{\mu,\varphi,\epsilon}(F_1 \cup F_2) + \eta.$$

The covering $(B(x_i, r_i))_i$ may be easily split into two parts,

$$I = \{\, i;\ B(x_i, r_i) \cap F_1 \neq \emptyset \,\} \quad \text{and} \quad J = \{\, i;\ B(x_i, r_i) \cap F_2 \neq \emptyset \,\},$$

for which $(B(x_i, r_i))_{i \in I}$, and $(B(x_i, r_i))_{i \in J}$ are centered δ-coverings of F_1, and F_2, respectively. As a result

$$\begin{aligned}
\overline{\mathcal{H}}^{\mathbf{q},t}_{\mu,\varphi,\epsilon}(F_1) + \overline{\mathcal{H}}^{\mathbf{q},t}_{\mu,\varphi,\epsilon}(F_2) &\leq \sum_{i \in I} (\mu(B(x_i, r_i)))^{\mathbf{q}} e^{t\varphi(r_i)} + \sum_{i \in J} \mu(B(x_i, r_i))^{\mathbf{q}} e^{t\varphi(r_i)} \\
&= \sum_i (\mu(B(x_i, r_i)))^{\mathbf{q}} e^{t\varphi(r_i)} \\
&\leq \overline{\mathcal{H}}^{q,t}_{\mu,\varphi,\epsilon}(F_1 \cup F_2) + \eta.
\end{aligned}$$

This implies that

$$\overline{\mathcal{H}}^{\mathbf{q},t}_{\mu,\varphi}(F_1) + \overline{\mathcal{H}}^{\mathbf{q},t}_{\mu,\varphi}(F_2) \leq \overline{\mathcal{H}}^{\mathbf{q},t}_{\mu,\varphi}(F_1 \cup F_2) + \epsilon \leq \mathcal{H}^{\mathbf{q},t}_{\mu,\varphi}(E_1 \cup E_2) + \eta.$$

So, as $\eta \downarrow 0$, we obtain

$$\overline{\mathcal{H}}^{\mathbf{q},t}_{\mu,\varphi}(F_1) + \overline{\mathcal{H}}^{\mathbf{q},t}_{\mu,\varphi}(F_2) \leq \overline{\mathcal{H}}^{\mathbf{q},t}_{\mu,\varphi}(F_1 \cup F_2) + \epsilon \leq \mathcal{H}^{\mathbf{q},t}_{\mu,\varphi}(E_1 \cup E_2).$$

Taking upper bound on all sets $F_1 \subseteq E_1$, and $F_2 \subseteq E_2$, we get

$$\mathcal{H}^{\mathbf{q},t}_{\mu,\varphi}(E_1 \cup E_2) \geq \mathcal{H}^{\mathbf{q},t}_{\mu,\varphi}(E_1) + \mathcal{H}^{\mathbf{q},t}_{\mu,\varphi}(E_2).$$

The opposite inequality is a natural consequence of the subadditivity of $\mathcal{H}^{\mathbf{q},t}_{\mu,\varphi}$.
3) We shall now show that $\mathcal{P}^{\mathbf{q},t}_{\mu,\varphi}$ is an outer measure on $\mathbb{R}^d$.
3.a) This is a natural consequence from the conctruction of $\mathcal{P}^{\mathbf{q},t}_{\mu,\varphi}$.
3.b) Let $E \subseteq F$ be two subsets of $\mathbb{R}^d$. We immediately observe that

$$\mathcal{P}^{\mathbf{q},t}_{\mu,\varphi}(E) = \inf_{E \subseteq \bigcup_i E_i} \sum_i \overline{\mathcal{P}}^{\mathbf{q},t}_{\mu,\varphi}(E_i) \leq \inf_{F \subseteq \bigcup_i E_i} \sum_i \overline{\mathcal{P}}^{\mathbf{q},t}_{\mu,\varphi}(E_i) = \mathcal{P}^{\mathbf{q},t}_{\mu,\varphi}(F).$$

3.c) Consider a sequence $(A_n)_n \subset \mathbb{R}^d$, $\eta > 0$, and let for each $n \in \mathbb{N}$, a covering $(E_{ni})_i$ of A_n, satisfying

$$\sum_i \overline{\mathcal{P}}^{\mathbf{q},t}_{\mu,\varphi}(E_{ni}) \leq \mathcal{P}^{\mathbf{q},t}_{\mu,\varphi}(A_n) + \frac{\eta}{2^n}.$$

It holds for all $\eta > 0$, that

$$\mathcal{P}^{\mathbf{q},t}_{\mu,\varphi}(\bigcup_n A_n) \leq \sum_n \sum_i \overline{\mathcal{P}}^{\mathbf{q},t}_{\mu,\varphi}(E_{ni}) \leq \sum_n \mathcal{P}^{\mathbf{q},t}_{\mu,\varphi}(A_n) + \eta.$$

As a result, we get

$$\mathcal{P}^{\mathbf{q},t}_{\mu,\varphi}(\bigcup_n A_n) \leq \sum_n \mathcal{P}^{\mathbf{q},t}_{\mu,\varphi}(A_n).$$

4) We will now prove that $\mathcal{P}^{\mathbf{q},t}_{\mu,\varphi}$ is metric on $\mathbb{R}^d$. The proof is more specific and uses the following result.

$$\overline{\mathcal{P}}^{\mathbf{q},t}_{\mu,\varphi}(A \cup B) = \overline{\mathcal{P}}^{\mathbf{q},t}_{\mu,\varphi}(A) + \overline{\mathcal{P}}^{\mathbf{q},t}_{\mu,\varphi}(B), \quad \text{whenever } d(A,B) > 0, \quad A, B \subset \mathbb{R}^d, \tag{7.2}$$

and where as previously, $d(.,.)$ is the Euclidean distance on $\mathbb{R}^d$. So, assuming that (7.2) is true, we deduce for all subsets $A, B \subset \mathbb{R}^d$, satisfying $d(A,B) > 0$, that whenever $0 < \epsilon < \frac{1}{2}d(A,B)$ and $(B(x_i,r_i))_i$ a centered ϵ-packing of the union set $A \cup B$, we may obtain as previously two parts I and J, satsifying

$$(B(x_i,r_i))_i = \Big(B(x_i,r_i)\Big)_{i\in I} \bigcup \Big(B(x_i,r_i)\Big)_{i\in J},$$

and

$$\forall i \in I, \ \ B(x_i,r_i) \cap B = \emptyset \quad \text{and} \quad \forall i \in J, \ \ B(x_i,r_i) \cap A = \emptyset.$$

This means that the part $(B(x_i,r_i))_{i\in I}$ constitutes a centered ϵ-packing of A and the second part $(B(x_i,r_i))_{i\in J}$ forms a centered ϵ-packing of B. As a consequence, we get

$$\sum_i (\mu(B(x_i,r_i)))^q e^{t\varphi(r_i)} = \underbrace{\sum_{i\in I} (\mu(B(x_i,r_i)))^q e^{t\varphi(r_i)}}_{\leq \overline{\mathcal{P}}^{\mathbf{q},t}_{\mu,\varphi,\epsilon}(A)} + \underbrace{\sum_{i\in I} (\mu(B(x_i,r_i)))^q e^{t\varphi(r_i)}}_{\leq \overline{\mathcal{P}}^{\mathbf{q},t}_{\mu,\varphi,\epsilon}(B)}.$$

Therefore,

$$\overline{\mathcal{P}}^{\mathbf{q},t}_{\mu,\varphi,\epsilon}(A \cup B) \leq \overline{\mathcal{P}}^{\mathbf{q},t}_{\mu,\varphi,\epsilon}(A) + \overline{\mathcal{P}}^{\mathbf{q},t}_{\mu,\varphi,\epsilon}(B).$$

Now, by letting $\epsilon \downarrow 0$, we obtain

$$\overline{\mathcal{P}}^{q,t}_{\mu,\varphi}(A \cup B) \leq \overline{\mathcal{P}}^{q,t}_{\mu,\varphi}(A) + \overline{\mathcal{P}}^{q,t}_{\mu,\varphi}(B).$$

Now to prove the converse, consider a centered ϵ-packing $(B(x_i,r_i))_i$ of A and a centered ϵ-packing $(B(y_i,r_i))_i$ of B. It is obvious that the union $\Big(B(x_i,r_i)\Big)_i \bigcup \Big(B(y_i,r_i)\Big)_i$ is a centered ϵ-packing of $A \cup B$. Therefore,

$$\overline{\mathcal{P}}^{\mathbf{q},t}_{\mu,\varphi,\epsilon}(A \cup B) \geq \sum_i (\mu(B(x_i,r_i)))^q e^{t\varphi(r_i)} + \sum_i (\mu(B(y_i,r_i)))^q e^{t\varphi(r_i)}.$$

Taking the upper bound on all the centered ϵ-packings $(B(x_i, r_i))_i$ of A, and similarly the upper bound on all the centered ϵ-packings $(B(y_i, r_i))_i$ of B, we deduce that

$$\overline{\mathcal{P}}^{q,t}_{\mu,\varphi,\epsilon}(A \cup B) \geq \overline{\mathcal{P}}^{q,t}_{\mu,\varphi,\epsilon}(A) + \overline{\mathcal{P}}^{q,t}_{\mu,\varphi,\epsilon}(B).$$

Now, as previously, letting $\epsilon \downarrow 0$, we obtain

$$\overline{\mathcal{P}}^{q,t}_{\mu,\varphi}(A \cup B) \geq \overline{\mathcal{P}}^{q,t}_{\mu,\varphi}(A) + \overline{\mathcal{P}}^{q,t}_{\mu,\varphi}(B).$$

Proof of (7.2). Recall that $\mathcal{P}^{q,t}_{\mu,\varphi}$ is an outer measure on $\mathbb{R}^d$. We thus may prove only the inequality

$$\mathcal{P}^{q,t}_{\mu,\varphi}(A \cup B) \geq \mathcal{P}^{q,t}_{\mu,\varphi}(A) + \mathcal{P}^{q,t}_{\mu,\varphi}(B).$$

Whenever the right-hand term is infinite, the result is obvious. So, we assume that $\mathcal{P}^{q,t}_{\mu,\varphi}(A \cup B) < \infty$. Let $\epsilon > 0$, and $(E_i)_i$ be a covering of the union $A \cup B$, satisfying

$$\sum_i \overline{\mathcal{P}}^{q,t}_{\mu,\varphi}(E_i) \leq \mathcal{P}^{q,t}_{\mu,\varphi}(A \cup B) + \epsilon.$$

Consider $F_i = A \cap E_i$, and $H_i = B \cap E_i$. It holds immediately that $(F_i)_i$, and $(H_i)_i$ are two coverings of A, and B, respectively that satisfy $F_i \cap H_j = \emptyset$ for all i, and j. We observe that

$$\mathcal{P}^{q,t}_{\mu,\varphi}(A) + \mathcal{P}^{q,t}_{\mu,\varphi}(B) \leq \sum_i (\overline{\mathcal{P}}^{q,t}_{\mu,\varphi}(F_i) + \overline{\mathcal{P}}^{q,t}_{\mu,\varphi}(H_i)).$$

As a result,

$$\overline{\mathcal{P}}^{q,t}_{\mu,\varphi}(A \cup B) = \overline{\mathcal{P}}^{q,t}_{\mu,\varphi}(A) + \overline{\mathcal{P}}^{q,t}_{\mu,\varphi}(B).$$

Therefore, we obtain

$$\mathcal{P}^{q,t}_{\mu,\varphi}(A) + \mathcal{P}^{q,t}_{\mu,\varphi}(B) \leq \sum_i \overline{\mathcal{P}}^{q,t}_{\mu,\varphi}(E_i) \leq \mathcal{P}^{q,t}_{\mu,\varphi}(A \cup B) + \epsilon.$$

It holds as for the case of the multifractal analysis of a single measure that the measures $\mathcal{H}^{q,t}_{\mu,\varphi}$, $\mathcal{P}^{q,t}_{\mu,\varphi}$ and the pre-measure $\overline{\mathcal{P}}^{q,t}_{\mu,\varphi}$ assign a dimension to every set $E \subseteq \mathbb{R}^d$. More precisely, the following result hold.

Proposition 7.1 *Given a subset $E \subseteq \mathbb{R}^d$,*

1. *There exists a unique number $dim^q_{\mu,\varphi}(E) \in [-\infty, +\infty]$ such that*

$$\mathcal{H}^{q,t}_{\mu,\varphi}(E) = \begin{cases} +\infty \text{ for } t < dim^q_{\mu,\varphi}(E) \\ 0 \quad si \ \ t > dim^q_{\mu,\varphi}(E) \end{cases}$$

2. *There exists a unique number $Dim^q_{\mu,\varphi}(E) \in [-\infty, +\infty]$ such that*

$$\mathcal{P}^{q,t}_{\mu,\varphi}(E) = \begin{cases} +\infty \text{ for } t < Dim^q_{\mu,\varphi}(E) \\ 0 \quad \text{for } t > Dim^q_{\mu,\varphi}(E) \end{cases}$$

3. There exists a unique number $\Delta^q_{\mu,\varphi}(E) \in [-\infty, +\infty]$ *such that*

$$\overline{\mathcal{P}}^{q,t}_{\mu,\varphi}(E) = \begin{cases} +\infty \text{ for } t < \Delta^q_{\mu,\varphi}(E) \\ 0 \ \text{ for } t > \Delta^q_{\mu,\varphi}(E) \end{cases}$$

Definition 7.2 For all set $E \subseteq \mathbb{R}^d$, we have

- $\dim^{\mathbf{q}}_{\mu,\varphi}(E)$ is called the φ-mixed multifractal generalization of the Hausdorff dimension of the set E.
- $\operatorname{Dim}^{\mathbf{q}}_{\mu,\varphi}(E)$ is called the φ-mixed multifractal generalization of the packing dimension of the set E.
- $\Delta^{\mathbf{q}}_{\mu,\varphi}(E)$ is called the φ-mixed multifractal generalization of the logarithmic index of the set E.

When $E = \operatorname{supp}(\mu)$, we denote respectively

$$b_{\mu,\varphi}(q) = \dim^{\mathbf{q}}_{\mu,\varphi}(E), \quad B_{\mu,\varphi}(q) = \operatorname{Dim}^{\mathbf{q}}_{\mu,\varphi}(E) \ \ \textit{and} \ \ \Delta_{\mu,\varphi}(q) = \Delta^{\mathbf{q}}_{\mu,\varphi}(E).$$

Remark that for $k = 1$ and φ the log function $\varphi(r) = \log(r)$, we come back to the classical definitions of the Hausdorff and packing measures and dimensions in their original forms (by taking $q = 0$) and their generalized multifractal variants for q being arbitrary. The mixed-case studied here may be also applied for a single measure and thus the results and characterizations outpointed in the present work remain valid for a single measure. Indeed, denote $Q_i = (0, 0, ..., q_i, 0, ..., 0)$ the vector with zero coordinates except for the ith one which equals q_i, we obtain the multifractal generalizations of the Hausdorff φ-measure and φ-dimension, the packing φ-dimension and the logarithmic φ-index of the set E for the single measure μ_i,

$$\dim^{Q_i}_{\mu,\varphi}(E) = \dim^{q_i}_{\mu_i,\varphi}(E),$$

$$\operatorname{Dim}^{Q_i}_{\mu,\varphi}(E) = \operatorname{Dim}^{q_i}_{\mu_i,\varphi}(E)$$

and

$$\Delta^{Q_i}_{\mu,\varphi}(E) = \Delta^{q_i}_{\mu_i,\varphi}(E).$$

Similarly, for the null vector of $\mathbb{R}^k$, we obtain

$$\dim^{0}_{\mu,\varphi}(E) = \dim_{\varphi}(E),$$

$$\operatorname{Dim}^{0}_{\mu,\varphi}(E) = \operatorname{Dim}_{\varphi}(E)$$

and

$$\Delta^{0}_{\mu,\varphi}(E) = \Delta_{\varphi}(E).$$

We may obtain further

$$\dim^{Q_i}_{\mu,\log}(E) = \dim^{q_i}_{\mu_i,\log}(E),$$

$$\operatorname{Dim}^{Q_i}_{\mu,\log}(E) = \operatorname{Dim}^{q_i}_{\mu_i,\log}(E)$$

and

$$\Delta^{Q_i}_{\mu,\log}(E) = \Delta^{q_i}_{\mu_i,\log}(E).$$

Similarly, for the null vector of $\mathbb{R}^k$, we obtain

$$\dim^{0}_{\mu,\log}(E) = \dim_{\log}(E) = \dim(E),$$

$$\mathrm{Dim}^0_{\mu,\log}(E) = \mathrm{Dim}_{\log}(E) = \mathrm{Dim}(E)$$

and

$$\Delta^0_{\mu,\log}(E) = \Delta_{\log}(E) = \Delta(E).$$

Proof of Proposition 7.1. We will sketch only the proof of the first point. The rest is analogous.
1. We claim that $\forall\, t \in \mathbb{R}$ such that $\mathcal{H}^{\mathbf{q},t}_{\mu,\varphi}(E) < \infty$ it holds that $\mathcal{H}^{\mathbf{q},t'}_{\mu,\varphi}(E) = 0$ for any $t' > t$. Indeed, let $\epsilon > 0$, $F \subseteq E$ and $(B(x_i, r_i))_i$ be a centered ϵ-covering of F. We have

$$\overline{\mathcal{H}}^{\mathbf{q},t'}_{\mu,\varphi,\epsilon}(F) \leq \sum_i (\mu(B(x_i,r_i)))^q e^{t'\varphi(r_i)} \leq e^{(t'-t)\varphi(\delta)} \sum_i (\mu(B(x_i,r_i)))^q e^{t\varphi(r_i)}.$$

Consequently,

$$\overline{H}^{\mathbf{q},t'}_{\mu,\varphi,\epsilon}(F) \leq e^{(t'-t)\varphi(\epsilon)} \overline{H}^{\mathbf{q},t}_{\mu,\varphi,\epsilon}(F).$$

Hence,

$$\overline{\mathcal{H}}^{\mathbf{q},t'}_{\mu,\varphi}(F) = 0, \quad \forall\, F \subseteq E.$$

As a result, $\mathcal{H}^{\mathbf{q},t'}_{\mu,\varphi}(E) = 0$. We then set

$$\dim^{\mathbf{q}}_{\mu,\varphi}(E) = \inf\{\, t \in \mathbb{R};\ \ \mathcal{H}^{\mathbf{q},t'}_{\mu,\varphi}(E) = 0\,\}.$$

One can proceed otherwise by claiming that $\forall\, t \in \mathbb{R}$ such that $\mathcal{H}^{\mathbf{q},t}_{\mu,\varphi}(E) > 0$ it holds that $\mathcal{H}^{\mathbf{q},t'}_{\mu,\varphi}(E) = +\infty$ for any $t' < t$. Indeed, proceeding as previously, we obtain for $\epsilon > 0$,

$$e^{(t'-t)\varphi(\epsilon)} \overline{H}^{\mathbf{q},t}_{\mu,\varphi,\epsilon}(F) \leq \overline{H}^{\mathbf{q},t'}_{\mu,\varphi,\epsilon}(F).$$

Hence,

$$\overline{\mathcal{H}}^{\mathbf{q},t'}_{\mu,\varphi}(F) = +\infty, \quad \forall\, F \subseteq E.$$

As a result, $\mathcal{H}^{\mathbf{q},t'}_{\mu,\varphi}(E) = +\infty$. We then set

$$\dim^{\mathbf{q}}_{\mu,\varphi}(E) = \sup\{\, t \in \mathbb{R};\ \ \mathcal{H}^{\mathbf{q},t'}_{\mu,\varphi}(E) = +\infty\,\}.$$

Next, we aim to study the characteristics of the mixed multifractal generalizations of dimensions.

Proposition 7.2

a. $b^{\mathbf{q}}_{\mu,\varphi}(.)$ *and* $B^{\mathbf{q}}_{\mu,\varphi}(.)$ *are non decreasing with respect to the inclusion property in* $\mathbb{R}^d$.
b. $b^{\mathbf{q}}_{\mu,\varphi}(.)$ *and* $B^{\mathbf{q}}_{\mu,\varphi}(.)$ *are* σ*-stable.*

Proof. **a.** Let $E \subseteq F$ be subsets of $\mathbb{R}^d$. We have

$$\mathcal{H}^{\mathbf{q},t}_{\mu,\varphi}(E) = \sup_{A \subseteq E} \overline{\mathcal{H}}^{\mathbf{q},t}_{\mu,\varphi}(A) \leq \sup_{A \subseteq F} \overline{\mathcal{H}}^{\mathbf{q},t}_{\mu,\varphi}(A) = \mathcal{H}^{\mathbf{q},t}_{\mu,\varphi}(F).$$

Which gives the monotony of $b^{\mathbf{q}}_{\mu,\varphi}(.)$.
b. Let $(A_n)_n$ be a countable set of subsets $A_n \subseteq \mathbb{R}^d$ and denote $A = \bigcup_n A_n$. It holds from the monotony of $b^{\mathbf{q}}_{\mu,\varphi}(.)$ that

$$b^{\mathbf{q}}_{\mu,\varphi}(A_n) \le b^{\mathbf{q}}_{\mu,\varphi}(A), \quad \forall n.$$

Hence,

$$\sup_n b^{\mathbf{q}}_{\mu,\varphi}(A_n) \le b^{\mathbf{q}}_{\mu,\varphi}(A).$$

Next, for any $t > \sup_n b^{\mathbf{q}}_{\mu,\varphi}(A_n)$, there holds that

$$\mathcal{H}^{\mathbf{q},t}_{\mu,\varphi}(A_n) = 0, \quad \forall n.$$

Consequently, from the sub-additivity property of $\mathcal{H}^{\mathbf{q},t}_{\mu,\varphi}$, it holds that

$$\mathcal{H}^{\mathbf{q},t}_{\mu,\varphi}(\bigcup_n A_n) = 0, \quad \forall t > \sup_n b^{\mathbf{q}}_{\mu,\varphi}(A_n).$$

Which means that

$$b^{\mathbf{q}}_{\mu,\varphi}(A) \le t, \quad \forall t > \sup_n b^{\mathbf{q}}_{\mu,\varphi}(A_n).$$

Hence,

$$b^{\mathbf{q}}_{\mu,\varphi}(A) \le \sup_n b^{\mathbf{q}}_{\mu,\varphi}(A_n).$$

Similar arguments permit us to prove the properties of $B^{\mathbf{q}}_{\mu,\varphi}(A)$. Next, we continue to study the characteristics of the mixed generalized multifractal dimensions. The following result is obtained.

Proposition 7.3

a. The functions $\boldsymbol{q} \longmapsto B_{\mu,\varphi}(\boldsymbol{q})$ and $\boldsymbol{q} \longmapsto \Lambda_{\mu,\varphi}(\boldsymbol{q})$ are convex.
b. For $i = 1, 2, \ldots, k$, the functions $q_i \longmapsto b_{\mu,\varphi}(\boldsymbol{q})$, $q_i \longmapsto B_{\mu,\varphi}(\boldsymbol{q})$ and $q_i \longmapsto \Lambda_{\mu,\varphi}(\boldsymbol{q})$ are non increasing.

Proof. a. We start by proving that $\Lambda_{\mu,\varphi}$ is convex. Let $\mathbf{p}, \mathbf{q} \in \mathbb{R}^k$, $\alpha \in]0, 1[$, $s > \Lambda_{\mu,\varphi}(\mathbf{p})$ and $t > \Lambda_{\mu,\varphi}(\mathbf{q})$. Consider next a centered ϵ-packing $(B_i = B(x_i, r_i))_i$ of E. Applying Hölder's inequality, it follows that

$$\sum_i (\mu(B_i))^{\alpha q+(1-\alpha)p} e^{(\alpha t+(1-\alpha)s)\varphi(r_i)} \le \left(\sum_i (\mu(B_i))^q e^{t\varphi(r_i)}\right)^{\alpha} \left(\sum_i (\mu(B_i))^p e^{s\varphi(r_i)}\right)^{1-\alpha}.$$

Hence,

$$\overline{\mathcal{P}}^{\alpha q+(1-\alpha)p,\alpha t+(1-\alpha)s}_{\mu,\varphi,\epsilon}(E) \le \left(\overline{\mathcal{P}}^{\mathbf{q},t}_{\mu,\varphi,\epsilon}(E)\right)^{\alpha} \left(\overline{\mathcal{P}}^{p,s}_{\mu,\varphi,\epsilon}(E)\right)^{1-\alpha}.$$

The limit on $\epsilon \downarrow 0$ gives

$$\overline{\mathcal{P}}^{\alpha q+(1-\alpha)p,\alpha t+(1-\alpha)s}_{\mu,\varphi}(E) \le \left(\overline{\mathcal{P}}^{\mathbf{q},t}_{\mu,\varphi}(E)\right)^{\alpha} \left(\overline{\mathcal{P}}^{\mathbf{p},s}_{\mu,\varphi}(E)\right)^{1-\alpha}.$$

Consequently,

$$\overline{\mathcal{P}}^{\alpha q+(1-\alpha)p,\alpha t+(1-\alpha)s}_{\mu,\varphi}(E) = 0 \ \ \forall s > \Lambda_{\mu,\varphi}(\mathbf{p}) \ \text{ and } \ t > \Lambda_{\mu,\varphi}(\mathbf{q}).$$

It results that

$$\Lambda_{\mu,\varphi}(\alpha\mathbf{q} + (1-\alpha)\mathbf{p}) \le \alpha\Lambda_{\mu,\varphi}(\mathbf{q}) + (1-\alpha)\Lambda_{\mu,\varphi}(\mathbf{p}).$$

We now prove the convexity of $B_{\mu,\varphi}$. We set in this case $t = B_{\mu,\varphi}(\mathbf{q})$ and $s = B_{\mu,\varphi}(\mathbf{p})$. We have

$$\mathcal{P}^{\mathbf{q},t+\varepsilon}_{\mu,\varphi}(E) = \mathcal{P}^{\mathbf{p},s+\varepsilon}_{\mu,\varphi}(E) = 0.$$

Therefore, there exists $(H_i)_i$ and $(K_i)_i$ coverings of the set E for which

$$\sum_i \overline{\mathcal{P}}^{\mathbf{q},t+\varepsilon}_{\mu,\varphi}(H_i) \leq 1 \qquad \text{et} \qquad \sum_i \overline{\mathcal{P}}^{\mathbf{p},s+\varepsilon}_{\mu,\varphi}(K_i) \leq 1.$$

Denote for $n \in \mathbb{N}$, $E_n = \bigcup_{1 \leq i,j \leq n} (H_i \cap K_j)$. Thus, $(E_n)_n$ is a covering of E. So that,

$$\begin{aligned}
&\mathcal{P}^{\alpha\mathbf{q}+(1-\alpha)\mathbf{p},\alpha t+(1-\alpha)s+\varepsilon}_{\mu,\varphi}(E_n) \\
&\leq \sum_{i,j=1}^{n} \mathcal{P}^{\alpha\mathbf{q}+(1-\alpha)\mathbf{p},\alpha t+(1-\alpha)s+\varepsilon}_{\mu,\varphi}(H_i \cap K_j) \\
&\leq \sum_{i,j=1}^{n} \overline{\mathcal{P}}^{\alpha\mathbf{q}+(1-\alpha)\mathbf{p},\alpha t+(1-\alpha)s+\varepsilon}_{\mu,\varphi}(H_i \cap K_j) \\
&\leq \left(\sum_{i,j=1}^{n} \overline{\mathcal{P}}^{\mathbf{q},t+\varepsilon}_{\mu,\varphi}(H_i \cap K_j)\right)^{\alpha} \left(\sum_{i,j=1}^{n} \overline{\mathcal{P}}^{\mathbf{p},s+\varepsilon}_{\mu,\varphi}(H_i \cap K_j)\right)^{1-\alpha} \\
&\leq n^{\alpha}\, n^{1-\alpha} = n < \infty.
\end{aligned}$$

Consequently,

$$B^{\alpha\mathbf{q}+(1-\alpha)\mathbf{p}}_{\mu,\varphi}(E_n) \leq \alpha\, t + (1-\alpha)s + \varepsilon, \quad \forall \varepsilon > 0.$$

Hence,

$$B^{\alpha\mathbf{q}+(1-\alpha)\mathbf{p}}_{\mu,\varphi}(E) \leq \alpha\, B^{\mathbf{q}}_{\mu,\varphi}(E) + (1-\alpha) B^{\mathbf{p}}_{\mu,\varphi}(E).$$

b. For $i = 1, 2, \ldots, k$, let $p_i \leq q_i$ reel numbers. Denote next

$$q = (q_1, \ldots, q_{i-1}, q_i, q_{i+1}, \ldots, q_k) \quad \text{and} \quad p = (q_1, \ldots, q_{i-1}, p_i, q_{i+1}, \ldots, q_k).$$

Let finally $A \subseteq E$. For a centered ϵ-covering $(B(x_i, r_i))_i$ of A, we have immediately

$$\mu(B(x_i, r_i))^{\mathbf{q}} e^{t\varphi(r_i)} \leq \mu(B(x_i, r_i))^{\mathbf{p}} e^{t\varphi(r_i)}, \quad \forall t \in \mathbb{R}.$$

Hence,

$$\overline{H}^{\mathbf{q},t}_{\mu,\varphi,\epsilon}(A) \leq \overline{H}^{\mathbf{p},t}_{\mu,\varphi,\epsilon}(A).$$

When $\epsilon \downarrow 0$, we obtain

$$\overline{H}^{\mathbf{q},t}_{\mu,\varphi}(A) \leq \overline{H}^{\mathbf{p},t}_{\mu,\varphi}(A).$$

Therefore,

$$\mathcal{H}^{\mathbf{q},t}_{\mu,\varphi}(E) = \sup_{A \subseteq E} \overline{\mathcal{H}}^{\mathbf{q},t}_{\mu,\varphi}(A) \leq \sup_{A \subseteq E} \overline{\mathcal{H}}^{\mathbf{p},t}_{\mu,\varphi}(A) = \mathcal{H}^{\mathbf{p},t}_{\mu,\varphi}(E).$$

This induces the fact that

$$\mathcal{H}^{\mathbf{q},t}_{\mu,\varphi}(E) = 0, \quad \forall t > b^{\mathbf{p}}_{\mu,\varphi}(E).$$

Consequently

$$b^{\mathbf{q}}_{\mu,\varphi}(E) < t, \quad \forall t > b^{\mathbf{p}}_{\mu,\varphi}(E).$$

Hence,

$$b^{\mathbf{q}}_{\mu,\varphi}(E) \leq b^{\mathbf{p}}_{\mu,\varphi}(E).$$

The remaining part to prove the monotony $\Lambda_{\mu,\varphi}$ and $B_{\mu,\varphi}$ is analogous.

Proposition 7.4 *The following inequality holds, for all* $(\boldsymbol{q}) \in \mathbb{R}^k$,

$$b_{\mu,\varphi}(\boldsymbol{q}) \leq B_{\mu,\varphi}(\boldsymbol{q}) \leq \Lambda_{\mu,\varphi}(\boldsymbol{q}).$$

The proof of this results reposes on the well known Besicovitch covering theorem (see Theorem 7.6) and the following intermediate Lemma.

Lemma 7.1 *There exists an integer* $\xi \geq 1$ *satisfying for any* $E \subseteq \mathbb{R}^d$,

$$\mathcal{H}^{q,t}_{\mu,\varphi}(E) \leq \xi \mathcal{P}^{q,t}_{\mu,\varphi}(E) \leq \xi \overline{\mathcal{P}}^{q,t}_{\mu,\varphi}(E), \; \forall \boldsymbol{q}, t.$$

More precisely, ξ *is the number related to the Besicovitch covering theorem.*

Proof. It suffices to prove the first inequality. The second is always true for all $\xi \geq 1$. Let $F \subseteq \mathbb{R}^d$, $\epsilon > 0$ and $\mathcal{V} = \{\, B(x, \frac{\epsilon}{2}); \quad x \in F\}$. Let next $\big((B_{ij})_j\big)_{1\leq i\leq \xi}$ be the ξ sets of $\mathcal{V}$ obtained by the Besicovitch covering theorem. So that, $(B_{ij})_{i,j}$ is a centered ϵ-covering of the set F and for each i, $(B_{ij})_j$ is a centered ϵ-packing of F. Therefore,

$$\overline{\mathcal{H}}^{\mathbf{q},t}_{\mu,\varphi,\epsilon}(F) \leq \sum_{i=1}^{\xi}\sum_{j}\big(\mu(B_{ij})\big)^{q} e^{t\varphi(r_{ij})} \leq \sum_{i=1}^{\xi}\overline{\mathcal{P}}^{\mathbf{q},t}_{\mu,\varphi,\epsilon}(F) = \xi \overline{\mathcal{P}}^{\mathbf{q},t}_{\mu,\varphi,\epsilon}(F).$$

Hence, $\overline{\mathcal{H}}^{\mathbf{q},t}_{\mu,\varphi}(F) \leq \xi \overline{\mathcal{P}}^{\mathbf{q},t}_{\mu,\varphi}(F)$. Consequently, for $E \subseteq \bigcup_i E_i$, we obtain

$$\begin{aligned}
\mathcal{H}^{\mathbf{q},t}_{\mu,\varphi}(E) = \mathcal{H}^{\mathbf{q},t}_{\mu,\varphi}\left(\bigcup_i (E_i \cap E)\right) &\leq \sum_i \mathcal{H}^{\mathbf{q},t}_{\mu,\varphi}(E_i \cap E)\\
&\leq \sum_i \sup_{F\subseteq E_i\cap E} \overline{\mathcal{H}}^{\mathbf{q},t}_{\mu,\varphi}(F)\\
&\leq \xi \sum_i \sup_{F\subseteq E_i\cap E} \overline{\mathcal{P}}^{\mathbf{q},t}_{\mu,\varphi}(F)\\
&\leq \xi \sum_i \overline{\mathcal{P}}^{\mathbf{q},t}_{\mu,\varphi}(E_i).
\end{aligned}$$

So as Lemma 7.1.

Proof of Proposition 7.4. It follows from Proposition 7.2, Proposition 7.3 and Lemma 7.1.

More details about such measures, dimensions, the associated multifractal formalism for non necessary Gibbs measures and also on applications and links to existing cases may be found in [82] and [112]. For example, for $k = 1$, $\varphi = \log$, $\mathbf{q} = 0$, we come back to the classical definitions of the Hausdorff and packing measures and dimensions in their original forms.

7.3 The φ-mixed Multifractal Densities

In this section we propose to develop our main results by introducing a mixed type of multifractal density for vector valued non necessary Gibbs measures relatively to the φ-mixed multifractal analysis developed in [82], [112]. Consider a vector valued measure $\mu = (\mu_1, \mu_2, \ldots, \mu_k)$ composed of probability measures on $\mathbb{R}^d$. For $x \in \mathrm{supp}(\mu)$, we define the upper and lower $(\mathbf{q}, t)$-densities of a probability measure ν with respect to μ and φ by

$$\overline{d}^{\mathbf{q},t}_{\mu,\varphi}(x,\nu) = \limsup_{r\to 0} \frac{\nu(B(x,r))}{\mu(B(x,r))^{\mathbf{q}} e^{t\varphi(r)}}$$

and

$$\underline{d}^{\mathbf{q},t}_{\mu,\varphi}(x,\nu) = \liminf_{r\to 0} \frac{\nu(B(x,r))}{\mu(B(x,r))^{\mathbf{q}} e^{t\varphi(r)}}.$$

For $a > 1$ and $1 \le j \le k$, we write

$$P^j_a(\mu) = \limsup_{r\searrow 0} \left(\sup_{x\in\mathrm{supp}(\mu_j)} \frac{\mu_j(B(x,ar))}{\mu_j(B(x,r))} \right).$$

We will now say that the measure μ satisfies the doubling condition if there exists $a > 1$ such that $P^j_a(\mu) < \infty$ for all $1 \le j \le k$. It is easily seen that the exact value of the parameter a is unimportant: $P^j_a(\mu) < \infty$, for some $a > 1$ if and only if $P^j_a(\mu) < \infty$, for all $a > 1$. Also, we will write $\mathcal{P}(\mathbb{R}^d)$ for the family of Borel probability measures on $\mathbb{R}^d$ and $\mathcal{P}_D(\mathbb{R}^d)$ for the family of Borel probability measures on $\mathbb{R}^d$ which satisfy the doubling condition. Our first main result which extends the results of [3], [44], [49], [33], [41], [42], [88], [90], [110] is stated as follows.

Theorem 7.2 *Let E be a Borel subset of* $\mathrm{supp}(\mu)$. *There exists a constant* $\xi > 0$ *such that*

1. *If* $\mathcal{H}^{q,t}_{\mu,\varphi}(E) < \infty$, *then*

$$\frac{1}{\xi}\mathcal{H}^{q,t}_{\mu,\varphi}(E) \inf_{x\in E} \overline{d}^{q,t}_{\mu,\varphi}(x,\nu) \le \nu(E) \le \mathcal{H}^{q,t}_{\mu,\varphi}(E) \sup_{x\in E} \overline{d}^{q,t}_{\mu,\varphi}(x,\nu), \tag{7.3}$$

2. *Let* φ *be a doubling function and* $\mu \in \mathcal{P}_D(\mathbb{R}^d)$. *If* $\mathcal{H}^{q,t}_{\mu,\varphi}(E) < \infty$, *then*

$$\mathcal{H}^{q,t}_{\mu,\varphi}(E) \inf_{x\in E} \overline{d}^{q,t}_{\mu,\varphi}(x,\nu) \le \nu(E) \le \mathcal{H}^{q,t}_{\mu,\varphi}(E) \sup_{x\in E} \overline{d}^{q,t}_{\mu,\varphi}(x,\nu). \tag{7.4}$$

3. *If* $\mathcal{P}^{q,t}_{\mu,\varphi}(E) < \infty$, *then*

$$\mathcal{P}^{q,t}_{\mu,\varphi}(E) \inf_{x\in E} \underline{d}^{q,t}_{\mu,\varphi}(x,\nu) \le \nu(E) \le \mathcal{P}^{q,t}_{\mu,\varphi}(E) \sup_{x\in E} \underline{d}^{q,t}_{\mu,\varphi}(x,\nu). \tag{7.5}$$

Proof of Theorem 7.2.
1. Denote $m = \inf_{x\in E} \overline{d}^{q,t}_{\mu,\varphi}(x,\nu)$. Without loss of generality we may assume that $m > 0$. Let $\varepsilon, \eta > 0$, and $F \subset E$ be closed set such that $\eta < m$. Let finally $H \subset F$ and denote for $\delta > 0$,

$$B_\delta(F) = \{x \in \mathbb{R}^d;\ dist(F,x) \le \delta\}.$$

As $B_F(\delta) \searrow F$ whenever $\delta \searrow 0$, there exists δ_0 for which

$$\nu\big(B_F(\delta)\big) \le \nu(F) + \frac{\epsilon}{2\xi}, \quad \forall\, 0 < \delta < \delta_0.$$

Let next $\delta < \delta_0$ be such that

$$\overline{\mathcal{H}}^{\mathbf{q},t}_{\mu,\varphi}(H) - \frac{\epsilon}{2(m-\eta)} \le \overline{\mathcal{H}}^{\mathbf{q},t}_{\mu,\varphi,\delta}(H).$$

and denote

$$\mathfrak{T}_\delta = \Big\{B(x,r), x \in H, 0 < r < \delta \text{ and } \nu\big(B(x,r)\big) \ge (m-\eta)\mu\big(B(x,r)\big)^{\mathbf{q}} e^{t\varphi(r)}\Big\}.$$

By applying Theorem 7.6, there exists a ξ countable or finite set $(\mathfrak{T}^i_\delta)_{1\le i\le \xi}$, with $\mathfrak{T}^i_\delta = \big(B(x_{ij}, r_{ij})\big)_j$, such that, for each i, $\mathfrak{T}^i_\delta$ consists of pairwise disjoint sets and $H \subset \bigcup_i \bigcup_{B\in\mathfrak{T}^i_\delta} B$. We then obtain

$$\begin{aligned}
\overline{\mathcal{H}}^{\mathbf{q},t}_{\mu,\varphi}(H)(m-\eta) &\le \overline{\mathcal{H}}^{\mathbf{q},t}_{\mu,\varphi,\delta}(H)(m-\eta) + \frac{\epsilon}{2}\\
&\le (m-\eta)\sum_{i=1}^{\xi}\sum_j \mu\big(B(x_{ij}, r_{ij})\big)^{\mathbf{q}} e^{t\varphi(r_{ij})} + \frac{\epsilon}{2}\\
&\le \sum_{i=1}^{\xi} \nu\big(\bigcup_j B(x_{ij}, r_{ij})\big) + \frac{\epsilon}{2}\\
&\le \xi\nu\big(B_F(\delta)\big) + \frac{\epsilon}{2}\\
&\le \nu(F) + \epsilon\\
&\le \nu(E) + \epsilon.
\end{aligned}$$

We now prove the second inequality of (7.3). Denote $M = \sup_{x\in E} \overline{d}^{\mathbf{q},t}_{\mu,\varphi}(x,\nu)$, $\varepsilon, \eta > 0$, $s \in \mathbb{N}$ and consider the set

$$E_s = \Big\{x \in E, \nu\big(B(x,r)\big) \le (M+\eta)\mu\big(B(x,r)\big)^{\mathbf{q}} e^{t\varphi(r)} \text{ and } 0 < r < \frac{1}{s}\Big\}.$$

There exists a $(1/s)$-centered covering $\big(B(x_i, r_i)\big)_i$ of E_s, such that

$$\sum_i \mu\big(B(x_i, r_i)\big)^{\mathbf{q}} e^{t\varphi(r_i)} \le \overline{\mathcal{H}}^{\mathbf{q},t}_{\mu,\varphi,\frac{1}{s}}(E_s) + \frac{\varepsilon}{M+\eta}.$$

Consequently,

$$\begin{aligned}
\nu(E_s) &\le \sum_i \nu\big(B(x_i, r_i)\big)\\
&\le (M+\eta)\sum_i \mu\big(B(x_i, r_i)\big)^{\mathbf{q}} e^{t\varphi(r_i)}\\
&\le (M+\eta)\overline{\mathcal{H}}^{\mathbf{q},t}_{\mu,\varphi,\frac{1}{s}}(E_s) + \varepsilon.
\end{aligned}$$

As $E_s \searrow E$ whenever $s \nearrow \infty$, we get

$$\nu(E) \leq (M+\eta)\overline{\mathcal{H}}^{\mathbf{q},t}_{\mu,\varphi}(E) + \varepsilon.$$

2. Let $m = \inf_{x\in E} \overline{d}^{\mathbf{q},t}_{\mu,\varphi}(x,\nu)$. Without loss of the generality we may assume that $m > 0$. Consider for $s \in \mathbb{N}^*$ the set

$$E_s = \left\{x \in E,\ \frac{\mu_j\big(B(x,5r)\big)}{\mu_j\big(B(x,r)\big)} < s,\ \forall\ 1 \leq j \leq k \text{ and } \frac{\varphi(5r)}{\varphi(r)} < s,\ 0 < r < \frac{1}{s}\right\}.$$

We claim as previously, that for any $\epsilon > 0, 0 < \eta < m$, $s \in \mathbb{N}^*$, for all closed subset $F \subset E_s$ and for any $H \subset F$, we have

$$\overline{\mathcal{H}}^{\mathbf{q},t}_{\mu,\varphi}(H)(m-\eta) \leq \nu(E) + \epsilon, \tag{7.6}$$

which in turns yields that

$$\mathcal{H}^{\mathbf{q},t}_{\mu,\varphi}(F)(m-\eta) \leq \nu(E) + \epsilon$$

and

$$\mathcal{H}^{\mathbf{q},t}_{\mu,\varphi}(E_s)(m-\eta) \leq \nu(E) + \epsilon.$$

Next, as $E_s \nearrow E$ when $s \nearrow \infty$ we get

$$\mathcal{H}^{\mathbf{q},t}_{\mu,\varphi}(E)(m-\eta) \leq \nu(E) + \epsilon.$$

It remains to prove the inequality (7.6). To do this, we consider for $\delta > 0$ the set

$$\mathfrak{T}_\delta = \left\{B(x,r), x \in H, 5r < \delta \text{ and } \nu\big(B(x,r)\big) \geq (m-\eta)\mu\big(B(x,r)\big)^{\mathbf{q}} e^{t\varphi(r)}\right\}.$$

Next let $\big(B(x_i,r_i)\big)_i$ be the countable disjoint sub-family of $\mathfrak{T}_\delta$ defined in Vitali's Theorem 7.8, such that

$$H \setminus \bigcup_{i=1}^{k} B(x_i,r_i) \subset \bigcup_{i\geq k} B(x_i,5r_i),\ \forall\, k \geq 1. \tag{7.7}$$

From the definition of E_s, there exists a constant $c = c(\mathbf{q},t,s) > 0$ such that

$$\begin{aligned}
\sum_i \mu\big(B(x_i,5r_i)\big)^{\mathbf{q}} e^{t\varphi(5r_i)} &\leq c \sum_i \mu\big(B(x_i,r_i)\big)^{\mathbf{q}} e^{t\varphi(r_i)} \\
&\leq c\,(m-\eta)^{-1} \sum_i \nu\big(B(x_i,r_i)\big) \\
&\leq c\,(m-\eta)^{-1} \nu\Big(\bigcup_i B(x_i,r_i)\Big) < \infty.
\end{aligned}$$

We deduce that there exists an integer N, such that

$$\sum_{i>N} \mu\big(B(x_i,5r_i)\big)^{\mathbf{q}} e^{t\varphi(5r_i)} \leq \frac{\epsilon}{3}(m-\eta)^{-1}.$$

Moreover, since $\overline{\mathcal{H}}^{\mathbf{q},t}_{\mu,\varphi}(H) < \infty$, we may choose $\delta < \delta_0$, with

$$\overline{\mathcal{H}}^{\mathbf{q},t}_{\mu,\varphi}(H) - \frac{\epsilon}{3(m-\eta)} \leq \overline{\mathcal{H}}^{\mathbf{q},t}_{\mu,\varphi,\delta}(H). \tag{7.8}$$

It follows from (7.7) and (7.8) that

$$\begin{aligned}\overline{\mathcal{H}}^{\mathsf{q},t}_{\mu,\varphi}(H)(m-\eta) &\leq \overline{\mathcal{H}}^{\mathsf{q},t}_{\mu,\varphi,\delta}(H)(m-\eta)+\frac{\epsilon}{3}\\ &\leq \sum_{i>N}\mu\big(B(x_i,5r_i)\big)^{\mathsf{q}}e^{t\varphi(5r_i)}+\sum_{i\leq N}\mu\big(B(x_i,r_i)\big)^{\mathsf{q}}e^{t\varphi(r_i)}+\frac{\epsilon}{3}\\ &\leq \sum_i \nu\big(B(x_i,r_i)\big)+\frac{2\epsilon}{3}.\end{aligned}$$

Recall now that when $\delta \searrow 0$, there holds that $B_F(\delta)\searrow F$. Consequently, there exists $\delta_0>0$ for which

$$\nu\big(B_F(\delta)\big)\leq \nu(F)+\frac{\epsilon}{3},\ \forall\, 0<\delta<\delta_0.$$

This implies that

$$\overline{\mathcal{H}}^{\mathsf{q},t}_{\mu,\varphi}(H)(m-\eta)\leq \nu\big(B_F(\delta)\big)+\frac{2\epsilon}{3}\leq \nu(E)+\epsilon.$$

3. We start by proving the right inequality of (7.5). Denote $a=\sup_{x\in E}\underline{d}^{\mathsf{q},t}_{\mu,\varphi}(x,\nu)$ and let $F\subset E$. It suffices to prove that

$$\nu(F)\leq a\overline{\mathcal{P}}^{\mathsf{q},t}_{\mu,\varphi}(F),\qquad \forall F\subset E. \tag{7.9}$$

Indeed, whenever (7.9) holds, we consider a covering $(E_i)_i$ of E and obtain

$$\begin{aligned}\nu(E)&=\nu\big(\cup_i(E\cap E_i)\big)\\ &\leq \sum_i \nu(E\cap E_i)\\ &\leq a\sum_i \overline{\mathcal{P}}^{\mathsf{q},t}_{\mu,\varphi}(E\cap E_i)\\ &\leq a\sum_i \overline{\mathcal{P}}^{\mathsf{q},t}_{\mu,\varphi}(E_i).\end{aligned}$$

Taking the inf over all the coverings $(E_i)_i$, the result follows immediately. We now proceed by proving (7.9). Let $F\subset E$, $\epsilon,\eta,\delta>0$ be such that

$$\overline{\mathcal{P}}^{\mathsf{q},t}_{\mu,\varphi,\delta}(F)\leq \overline{\mathcal{P}}^{\mathsf{q},t}_{\mu,\varphi}(F)+\frac{\epsilon}{a+\eta}.$$

Consider next the set

$$\mathfrak{T}_\delta=\Big\{B(x,r),x\in F,r<\delta \ \text{ and } \ \nu\big(B(x,r)\big)\leq (a+\eta)\mu\big(B(x,r)\big)^{\mathsf{q}}e^{t\varphi(r)}\Big\}.$$

By Theorem 7.7, there exists a δ-packing $\big(B(x_i,r_i)\big)_i\subset\mathfrak{T}_\delta$ of F satisfying

$$\nu\big(F\setminus \cup_i B(x_i,r_i)\big)=0.$$

Moreover, we have

$$\begin{aligned}\nu(F) &= \nu\big(\cup_i\,(F\cap B(x_i,r_i))\big)\\ &\le \sum_i \nu(B(x_i,r_i))\\ &\le (a+\eta)\sum_i \mu\big(B(x_i,r_i)\big)^{\mathsf{q}} e^{t\varphi(r_i)}\\ &\le (a+\eta)\overline{\mathcal{P}}^{\mathsf{q},t}_{\mu,\varphi,\delta}(F)\\ &\le (a+\eta)\overline{\mathcal{P}}^{\mathsf{q},t}_{\mu,\varphi}(F)+\epsilon.\end{aligned}$$

Letting $\epsilon \to 0$, we obtain

$$\nu(F) \le (a+\eta)\overline{\mathcal{P}}^{\mathsf{q},t}_{\mu,\varphi}(F).$$

Next, making $\eta \to 0$ equation (7.9) follows immediately.
We now proceed by proving the left inequality of (7.5). Denote $m = \inf_{x\in E} \underline{d}^{\mathsf{q},t}_{\mu,\varphi}(x,\nu)$ and assume with out loss of the generality that $m > 0$. We just need to prove that

$$\mathcal{P}^{\mathsf{q},t}_{\mu,\varphi}(E)(m-\eta) \le \nu(E)+\varepsilon, \qquad \forall\, \varepsilon>0, \quad \forall\, 0<\eta<m.$$

Fix $\varepsilon > 0$ and $0 < \eta < m$. It is sufficient to prove that, for any closed subset F of E,

$$\mathcal{P}^{\mathsf{q},t}_{\mu,\varphi}(F)(m-\eta) \le \nu(E)+\varepsilon.$$

Recall here again that, if $\delta \searrow 0$, then $B_F(\delta) \searrow F$. So, there exists δ_0 satisfying

$$\nu\big(B_F(\delta)\big) \le \nu(F)+\epsilon, \qquad \forall\, 0<\delta<\delta_0.$$

For $s \in \mathbb{N}$, consider the set

$$F_s = \Big\{x\in F,\, \nu\big(B(x,r)\big) \ge (m-\eta)\mu\big(B(x,r)\big)^{\mathsf{q}} e^{t\varphi(r)}, \ \text{ for } \ 0<r<\frac{1}{s}\Big\}.$$

Fix $s \in \mathbb{N}$ and $0 < \delta < \sup\{\frac{1}{s}, \delta_0\}$. Let $\big(B(x_i,r_i)\big)_i$ be a centered δ-packing of F_s. Then,

$$\begin{aligned}(m-\eta)\sum_i \mu(B(x_i,r_i))^{\mathsf{q}} e^{t\varphi(r_i)} &\le \sum_i \nu\big(B(x_i,r_i)\big)\\ &= \nu\big(\cup_i\, B(x_i,r_i)\big)\\ &\le \nu\big(B_F(\delta)\big)\\ &\le \nu(F)+\epsilon\\ &\le \nu(E)+\epsilon.\end{aligned}$$

Hence,

$$\mathcal{P}^{\mathsf{q},t}_{\mu,\varphi}(F_s)(m-\eta) \le \overline{\mathcal{P}}^{\mathsf{q},t}_{\mu,\varphi}(F_s)(m-\eta) \le \overline{\mathcal{P}}^{\mathsf{q},t}_{\mu,\varphi,\delta}(F_s)(m-\eta) \le \nu(E)+\varepsilon.$$

As $F_s \nearrow F$ when $s \nearrow \infty$, we obtain

$$\mathcal{P}^{\mathsf{q},t}_{\mu,\varphi}(F)(m-\eta) \le \nu(E)+\varepsilon.$$

In the following part we propose to link the previous estimations of the φ-mixed multifractal densities to the exact computation of the both φ-mixed generaliations of Hausdorff and packing measures and dimensions. We will show precisely that these densities permit in some special cases to compute the φ-mixed multifractal generalizations of both Hausdorff and packing dimensions of sets characterized by the existence of some suitable measure(s) supported on them. This problem has been the object of several papers such as [3], [33], [49], ...

Let $E \subset \mathbb{R}^d$ be a Borel subset and denote by $\mathcal{H}^{\mathbf{q},s}_{\mu,\varphi} \llcorner E$ (resp. $\mathcal{P}^{\mathbf{q},t}_{\mu,\varphi} \llcorner E$) the s-dimensional centered Hausdorff measure $\mathcal{H}^{\mathbf{q},s}_{\mu,\varphi}$ (resp. t-dimensional centered packing measure $\mathcal{P}^{\mathbf{q},t}_{\mu,\varphi}$) restricted to E (if the Hausdorff or packing measure of E is zero, then the restriction measure is in fact a zero measure).

For $\nu = \mathcal{H}^{\mathbf{q},t}_{\mu,\varphi} \llcorner E$, we define

$$\overline{D}^{\mathbf{q},t}_{\mu,\varphi}(x, E) = \overline{d}^{\mathbf{q},t}_{\mu,\varphi}(x, \nu) \qquad \text{and} \qquad \underline{D}^{\mathbf{q},t}_{\mu,\varphi}(x, E) = \underline{d}^{\mathbf{q},t}_{\mu,\varphi}(x, \nu).$$

Similarly, for $\nu = \mathcal{P}^{\mathbf{q},t}_{\mu,\varphi} \llcorner E$, we define

$$\overline{\Delta}^{\mathbf{q},t}_{\mu,\varphi}(x, E) = \overline{d}^{q,t}_{\mu,\varphi}(x, \nu) \qquad \text{and} \qquad \underline{\Delta}^{\mathbf{q},t}_{\mu,\varphi}(x, E) = \underline{d}^{\mathbf{q},t}_{\mu,\varphi}(x, \nu).$$

Whenever $\overline{D}^{\mathbf{q},t}_{\mu,\varphi}(x, E) = \underline{D}^{\mathbf{q},t}_{\mu,\varphi}(x, E)$ (resp. $\overline{\Delta}^{\mathbf{q},t}_{\mu,\varphi}(x, \nu) = \underline{\Delta}^{\mathbf{q},t}_{\mu,\varphi}(x, \nu)$), we write $D^{\mathbf{q},t}_{\mu,\varphi}(x, E)$ (resp. $\Delta^{\mathbf{q},t}_{\mu,\varphi}(x, E)$) for the common value.

As a result of Theorem 7.2 we obtain the estimations

$$\inf_{x\in E} \overline{D}^{\mathbf{q},t}_{\mu,\varphi}(x, E) \le 1 \le \sup_{x\in E} \overline{D}^{\mathbf{q},t}_{\mu,\varphi}(x, E)$$

and

$$\inf_{x\in E} \underline{\Delta}^{q,t}_{\mu,\varphi}(x, E) \le 1 \le \sup_{x\in E} \underline{\Delta}^{\mathbf{q},t}_{\mu,\varphi}(x, E).$$

An interesting question is then to study the case of equality for the last inequalities. The equality means in some sense that the measure ν plays the role of the Gibbs measure constructed on the singularities set of the vector valued measure μ. Here, it means that the vector valued measure μ when controled by the gauge function φ permits the construction of a Borel probability measure ν on the φ-mixed singularities sets which permit in turns to prove the validity of an associated variant of the multifractal formalism conjectured in [82]. So, consider the sets

$$\underline{K} = \left\{x \in E,\ \underline{D}^{\mathbf{q},t}_{\mu,\varphi}(x, E) = 1\right\},\ \overline{K} = \left\{x \in E,\ \overline{D}^{\mathbf{q},t}_{\mu,\varphi}(x, E) = 1\right\},$$

$$\underline{T} = \left\{x \in E,\ \underline{\Delta}^{\mathbf{q},t}_{\mu,\varphi}(x, E) = 1\right\},\ \overline{T} = \left\{x \in E,\ \overline{\Delta}^{\mathbf{q},t}_{\mu,\varphi}(x, E) = 1\right\},$$

$$K = \underline{K} \cap \overline{K} \quad \text{and} \quad T = \underline{T} \cap \overline{T}.$$

The following result provides a description of these sets by means of their φ-mixed multifractal generalizations of Hausdorff and packing dimensions.

Theorem 7.3 *Let E be a Borel subset of* $\operatorname{supp}(\mu)$.

1. *Let φ be a doubling function and $\mu \in \mathcal{P}_D(\mathbb{R}^d)$. If $\mathcal{H}^{q,t}_{\mu,\varphi}(E) < \infty$, then* $\dim^{q}_{\mu,\varphi}(\overline{K}) = t$.
2. *If $\mathcal{P}^{q,t}_{\mu,\varphi}(E) < \infty$, then* $\operatorname{Dim}^{q}_{\mu,\varphi}(\underline{T}) = t$.

3. *Let φ be a doubling function and $\mu \in \mathcal{P}_D(\mathbb{R}^d)$. If $\mathcal{P}^{q,t}_{\mu,\varphi}(E) < \infty$, then the following assertions are equivalent*

   a. $\mathcal{H}^{q,t}_{\mu,\varphi} = \mathcal{P}^{q,t}_{\mu,\varphi}$.
   b. $\overline{D}^{q,t}_{\mu,\varphi}(x, E) = 1 = \underline{D}^{q,t}_{\mu,\varphi}(x, E)$ *for* $\mathcal{P}^{q,t}_{\mu,\varphi}$*-a.a.* $x \in E$.
   c. $\overline{\Delta}^{q,t}_{\mu,\varphi}(x, E) = 1 = \underline{\Delta}^{q,t}_{\mu,\varphi}(x, E)$ *for* $\mathcal{P}^{q,t}_{\mu,\varphi}$*-a.a.* $x \in E$.

4. *If $\mathcal{H}^{q,t}_{\mu,\varphi} = \mathcal{P}^{q,t}_{\mu,\varphi} < \infty$, then*

$$\dim^q_{\mu,\varphi}(\overline{K}) = \dim^q_{\mu,\varphi}(K) = \dim^q_{\mu,\varphi}(\overline{T}) = \dim^q_{\mu,\varphi}(T) = t.$$

This result is important as it consists of a first information leading to the computation of the multifractal spectrum due to the introduced densities. Indeed, related to the original form of the multifractal spectrum, the starting point is to establish an estimation of the form

$$\nu(B(x,r)) \sim (\mu(B(x,r))^q e^{(t\pm\varepsilon)\varphi(r)}, \quad r \to 0.$$

For example, when φ is the classical logarithm ($\varphi(r) = \log r$), we obtain an estimation of the form

$$\nu(B(x,r)) \sim (\mu(B(x,r))^q (2r)^{t\pm\varepsilon}, \quad r \to 0$$

which in the case of a Hölderian (Gibbs) measur μ means that the densities considered above are all equals 1 and thus permits to compute the multifractal spectrum (evaluated as the Hausdorff dimension of the level sets of the densities) by means of a Legendre transform of a convex function issued from the multifractal generalized dimensions $b^q_{\mu,\varphi}$, $B^q_{\mu,\varphi}$ and $\Delta^q_{\mu,\varphi}$.

Proof of Theorem 8.7.

1. We show firstly that

$$\overline{D}^{q,t}_{\mu,\varphi}(x, E) \le 1, \quad \text{for } \mathcal{H}^{q,t}_{\mu,\varphi}\text{-a.a. } x \in E. \tag{7.10}$$

Consider the sets

$$F = \left\{x \in E,\ \overline{D}^{q,t}_{\mu,\varphi}(x, E) > 1\right\}$$

and

$$F_m = \left\{x \in E,\ \overline{D}^{q,t}_{\mu,\varphi}(x, E) > 1 + \frac{1}{m}\right\}, \quad m \in \mathbb{N}.$$

It follows from (7.4) that

$$\left(1 + \frac{1}{m}\right)\mathcal{H}^{q,t}_{\mu,\varphi}(F_m) \le \mathcal{H}^{q,t}_{\mu,\varphi}(F_m).$$

Consequently, $\mathcal{H}^{q,t}_{\mu,\varphi}(F_m) = 0$. Since $F = \underset{m}{\cup} F_m$, we obtain (7.10).

We now prove the inequality

$$1 \le \overline{D}^{q,t}_{\mu,\varphi}(x, E) \quad \text{for } \mathcal{H}^{q,t}_{\mu,\varphi}\text{-a.a. } x \in E. \tag{7.11}$$

We consider as previously the sets

$$G = \left\{x \in E,\ \overline{D}^{q,t}_{\mu,\varphi}(x, E) < 1\right\}$$

and

$$G_m = \left\{x \in E,\ \overline{D}^{\mathbf{q},t}_{\mu,\varphi}(x, E) \leq 1 - \frac{1}{m}\right\},\quad m \in \mathbb{N}.$$

Applying (7.4) we obtain

$$\left(1 - \frac{1}{m}\right)\mathcal{H}^{\mathbf{q},t}_{\mu,\varphi}(G_m) \geq \mathcal{H}^{\mathbf{q},t}_{\mu,\varphi}(G_m).$$

Then, $\mathcal{H}^{\mathbf{q},t}_{\mu,\varphi}(G_m) = 0$. Since $G = \underset{m}{\cup} G_m$, we obtain (7.11).
Finally, (7.10) and (7.11) lead to the desired result.
2. The proof is similar to assertion **1**.
3. $\underline{(a) \Rightarrow (b)}$. It follows from assertion **1** that for $\mu \in \mathcal{P}_D(E)$ and $\mathcal{H}^{\mathbf{q},t}_{\mu,\varphi}(E) < \infty$. Thus

$$\overline{D}^{\mathbf{q},t}_{\mu,\varphi}(x, E) = 1 \quad \text{for} \quad \mathcal{H}^{\mathbf{q},t}_{\mu,\varphi}\text{-a.a. } x \in E. \tag{7.12}$$

Next, observe that

$$\mathcal{H}^{\mathbf{q},t}_{\mu,\varphi}(F) = \mathcal{P}^{\mathbf{q},t}_{\mu,\varphi}(F), \quad \text{for any} \quad F \subset E. \tag{7.13}$$

Thanks to (7.12) and (7.13), we get

$$\overline{D}^{\mathbf{q},t}_{\mu,\varphi}(x, E) = 1 \quad \text{for} \quad \mathcal{P}^{\mathbf{q},t}_{\mu,\varphi}\text{-a.a. } x \in E. \tag{7.14}$$

Now, we consider the sets

$$F = \left\{x \in E, \underline{D}^{\mathbf{q},t}_{\mu,\varphi}(x, E) < 1\right\},$$

and

$$F_m = \left\{x \in E, \underline{D}^{\mathbf{q},t}_{\mu,\varphi}(x, E) < 1 - \frac{1}{m}\right\},\quad m \in \mathbb{N}.$$

From Theorem 7.2 and (7.14), we obtain

$$\mathcal{P}^{\mathbf{q},t}_{\mu,\varphi}(F_m) = \mathcal{H}^{\mathbf{q},t}_{\mu,\varphi}(F_m) \leq \left(1 - \frac{1}{m}\right)\mathcal{P}^{\mathbf{q},t}_{\mu,\varphi}(F_m).$$

This implies that $\mathcal{P}^{\mathbf{q},t}_{\mu,\varphi}(F_m) = 0$. As $F = \underset{m}{\cup} F_m$, we obtain $\mathcal{P}^{\mathbf{q},t}_{\mu,\varphi}(F) = 0$, i.e.,

$$\underline{D}^{\mathbf{q},t}_{\mu,\varphi}(x, E) \geq 1 \quad \text{for} \quad \mathcal{P}^{\mathbf{q},t}_{\mu,\varphi}\text{-a.a. } x \in E. \tag{7.15}$$

Finally, (7.14) and (7.15) lead to (b).
$\underline{(b) \Rightarrow (a)}$. Consider the set

$$F = \left\{x \in E, D^{\mathbf{q},t}_{\mu,\varphi}(x, E) = 1\right\}.$$

It follows from Theorem 7.2 and (b) that

$$\mathcal{P}^{\mathbf{q},t}_{\mu,\varphi}(E) = \mathcal{P}^{\mathbf{q},t}_{\mu,\varphi}(F) \leq \mathcal{H}^{\mathbf{q},t}_{\mu,\varphi}(F) \leq \mathcal{H}^{\mathbf{q},t}_{\mu,\varphi}(E) \leq \mathcal{P}^{\mathbf{q},t}_{\mu,\varphi}(E),$$

which yields that $\mathcal{H}^{q,t}_{\mu,\varphi}(E) = \mathcal{P}^{q,t}_{\mu,\varphi}(E)$.
$(b) \Leftrightarrow (c)$. It may be checked by following similar techniques as in the proof of $\Big((a) \Leftrightarrow (b)\Big)$.
4. It is an immediate consequence of **3**.

7.4 Regularities of φ-mixed Multifractal Measures

In this section, we will prove a decomposition theorem of Besicovitch's type for the φ-mixed multifractal generalizations of Hausdorff and packing measures.

Definition 7.3 Let $E \subset \text{supp}(\mu)$ be a Borel set with $0 < \mathcal{H}^{q,t}_{\mu,\varphi}(E) < +\infty$. For backgrounds the readers may refer to [2], [104], [108], [109].

1. A point $x \in E$ is called $\mathcal{H}^{q,t}_{\mu,\varphi}$-regular point of E if $D^{q,t}_{\mu,\varphi}(x, E)$ exists. Otherwise x is a $\mathcal{H}^{q,t}_{\mu,\varphi}$-irregular point.
2. E is said to be $\mathcal{H}^{q,t}_{\mu,\varphi}$-regular if $\mathcal{H}^{q,t}_{\mu,\varphi}$-almost all its points are $\mathcal{H}^{q,t}_{\mu,\varphi}$-regular.
3. E is said to be $\mathcal{H}^{q,t}_{\mu,\varphi}$-irregular if $\mathcal{H}^{q,t}_{\mu,\varphi}$-almost all its points are $\mathcal{H}^{q,t}_{\mu,\varphi}$-irregular.

Remark 7.1 Similarly, we define the regularities for the relative φ-mixed multifractal packing measure $\mathcal{P}^{q,t}_{\mu,\varphi}$ by replacing $\mathcal{H}^{q,t}_{\mu,\varphi}$ in Definition 7.3 above by $\mathcal{P}^{q,t}_{\mu,\varphi}$.

Lemma 7.2 *Let $E \subset \text{supp}(\mu)$ be a Borel subset and $F \subseteq E$ be $\mathcal{H}^{q,t}_{\mu,\varphi}$-measurable.*

1. Whenever $\mathcal{H}^{q,t}_{\mu,\varphi}(E) < \infty$ we have

$$\overline{D}^{q,t}_{\mu,\varphi}(x, E) = \overline{D}^{q,t}_{\mu,\varphi}(x, F) \text{ and } \underline{D}^{q,t}_{\mu,\varphi}(x, E) = \underline{D}^{q,t}_{\mu,\varphi}(x, F), \text{ for } \mathcal{H}^{q,t}_{\mu,\varphi}\text{-a.e. } x \in F.$$

2. Whenever $\mathcal{P}^{q,t}_{\mu,\varphi}(E) < \infty$ we have

$$\overline{\Delta}^{q,t}_{\mu,\varphi}(x, E) = \overline{\Delta}^{q,t}_{\mu,\varphi}(x, F) \text{ and } \underline{\Delta}^{q,t}_{\mu,\varphi}(x, E) = \underline{\Delta}^{q,t}_{\mu,\varphi}(x, F), \text{ for } \mathcal{H}^{q,t}_{\mu,\varphi}\text{-a.e. } x \in F.$$

Proof. Let $\nu \in \mathcal{P}(\mathbb{R}^d)$ and define the measure ν_F by

$$\nu_F(A) = \nu(F \cap A),$$

for all Borel set A. Assume that $\mathcal{H}^{q,t}_{\mu,\varphi}(E) < \infty$. We will prove that for $\mathcal{H}^{q,t}_{\mu,\varphi}$-a.e. $x \in F$,

$$\overline{d}^{q,t}_{\mu,\varphi}(x, \nu) = \overline{d}^{q,t}_{\mu,\varphi}(x, \nu_F) \text{ and } \underline{d}^{q,t}_{\mu,\varphi}(x, \nu) = \underline{d}^{q,t}_{\mu,\varphi}(x, \nu_F). \tag{7.16}$$

Indeed, we already now that

$$\underline{d}^{q,t}_{\mu,\varphi}(x, \nu) \geq \underline{d}^{q,t}_{\mu,\varphi}(x, \nu_F) \qquad \text{and} \qquad \overline{d}^{q,t}_{\mu,\varphi}(x, \nu) \geq \overline{d}^{q,t}_{\mu,\varphi}(x, \nu_F).$$

Denote next $\lambda(A) = \nu(A \setminus F)$ for all $A \subseteq X$. Then,

$$\nu(A) = \nu\big(A \cap (F^c \cup F)\big) = \nu(A \setminus F) + \nu(A \cap F) = \lambda(A) + \nu_F(A).$$

It holds that

$$\underline{d}^{\mathbf{q},t}_{\mu,\varphi}(x,\nu) \le \underline{d}^{\mathbf{q},t}_{\mu,\varphi}(x,\nu_F) + \overline{d}^{\mathbf{q},t}_{\mu,\varphi}(x,\lambda) \text{ and } \overline{d}^{\mathbf{q},t}_{\mu,\varphi}(x,\nu) \le \overline{d}^{\mathbf{q},t}_{\mu,\varphi}(x,\nu_F) + \overline{d}^{\mathbf{q},t}_{\mu,\varphi}(x,\lambda).$$

Consequently, it suffices to show that $\overline{d}^{\mathbf{q},t}_{\mu,\varphi}(x,\lambda) = 0$. So, for $k \in \mathbb{N}$, let

$$F_k = \left\{ x \in F;\ \overline{d}^{\mathbf{q},t}_{\mu,\varphi}(x,\lambda) \ge \frac{1}{k} \right\}.$$

By (7.3), we immediately conclude that

$$\frac{1}{k\xi}\mathcal{H}^{\mathbf{q},t}_{\mu,\varphi}(F_k) \le \lambda(F_k) = \nu(F_k \setminus F) = \nu(\emptyset) = 0, \qquad \text{for all } k \ge 1,$$

which yields that $\overline{d}^{\mathbf{q},t}_{\mu,\varphi}(x,\lambda) = 0$, for $\mathcal{H}^{\mathbf{q},t}_{\mu,\varphi}$-a.e. $x \in F$ and thus leads to (7.16).
Now, in (7.16), taking $\nu = \mathcal{H}^{\mathbf{q},t}_{\mu,\varphi}{}_{\llcorner E}$ $\left(\text{resp. } \nu = \mathcal{P}^{\mathbf{q},t}_{\mu,\varphi}{}_{\llcorner E}\right)$, we obtain Assertion **1** (resp. Assertion **2**) of the Lemma.

Lemma 7.3 *Let E be a Borel subset of* supp(μ) *with $\mathcal{P}^{q,t}_{\mu,\varphi}(E) < \infty$ and $F = \left\{x \in E;\ \overline{\Delta}^{q,t}_{\mu,\varphi}(x,E) < +\infty\right\}$. If G is a Borel subset of F such that $\mathcal{H}^{q,t}_{\mu,\varphi}(G) = 0$, then $\mathcal{P}^{q,t}_{\mu,\varphi}(G) = 0$.*

Proof. It follows from (7.3) by taking $\nu = \mathcal{P}^{\mathbf{q},t}_{\mu,\varphi}{}_{\llcorner E}$.

Theorem 7.4 *Let E be a Borel subset of* supp(μ), *φ satisfying (7.1) and $\mu \in \mathcal{P}_D(\mathbb{R}^d)$.*

1. *If $\mathcal{H}^{q,t}_{\mu,\varphi}(E) < +\infty$, then the set of $\mathcal{H}^{q,t}_{\mu,\varphi}$-regular points of E is $\mathcal{H}^{q,t}_{\mu,\varphi}$-regular and the set of $\mathcal{H}^{q,t}_{\mu,\varphi}$-irregular points of E is $\mathcal{H}^{q,t}_{\mu,\varphi}$-irregular.*
2. *If $\mathcal{P}^{q,t}_{\mu,\varphi}(E) < +\infty$, then the set of $\mathcal{P}^{q,t}_{\mu,\varphi}$-regular points of E is $\mathcal{P}^{q,t}_{\mu,\varphi}$-regular and the set of $\mathcal{P}^{q,t}_{\mu,\varphi}$-irregular points of E is $\mathcal{P}^{q,t}_{\mu,\varphi}$-irregular.*

Proof.

1. Put $F = \left\{x \in E;\ D^{\mathbf{q},t}_{\mu,\varphi}(x,E) = 1\right\}$. Since $F \subset E$ and $\mathcal{H}^{\mathbf{q},t}_{\mu,\varphi}(E) < \infty$, from Theorem 7.3, we have $\overline{D}^{\mathbf{q},t}_{\mu,\varphi}(x,F) = 1$, for $\mathcal{H}^{\mathbf{q},t}_{\mu,\varphi}$-a.e. $x \in F$. So, we only have to prove that $\underline{D}^{\mathbf{q},t}_{\mu,\varphi}(x,F) = 1$, for $\mathcal{H}^{\mathbf{q},t}_{\mu,\varphi}$-a.e. $x \in F$. Lemma 7.2 implies that

   $$\underline{D}^{\mathbf{q},t}_{\mu,\varphi}(x,F) = \underline{D}^{\mathbf{q},t}_{\mu,\varphi}(x,E), \quad \text{for } \mathcal{H}^{\mathbf{q},t}_{\mu,\varphi}\text{-a.e. } x \in F.$$

   We therefore conclude that $\underline{D}^{\mathbf{q},t}_{\mu,\varphi}(x,F) = 1$, for $\mathcal{H}^{\mathbf{q},t}_{\mu,\varphi}$-a.e. $x \in F$. Again, by Lemma 7.2,

   $$\overline{D}^{\mathbf{q},t}_{\mu,\varphi}(x,E \setminus F) = \overline{D}^{\mathbf{q},t}_{\mu,\varphi}(x,E), \quad \text{for } \mathcal{H}^{\mathbf{q},t}_{\mu,\varphi}\text{-a.e. } x \in E \setminus F$$

   and

   $$\underline{D}^{\mathbf{q},t}_{\mu,\varphi}(x,E \setminus F) = \underline{D}^{\mathbf{q},t}_{\mu,\varphi}(x,E), \quad \text{for } \mathcal{H}^{\mathbf{q},t}_{\mu,\varphi}\text{-a.e. } x \in E \setminus F.$$

   Finally, it follows that

   $$\mathcal{H}^{\mathbf{q},t}_{\mu,\varphi}\Big(\left\{x \in E \setminus F,\ D^{\mathbf{q},t}_{\mu,\varphi}(x,E \setminus F) = 1\right\}\Big) = 0.$$
2. The proof is similar to assertion 1.

Now, in the remaining part of this section, we propose to establish a necessary condition for a strong regularity with the φ-mixed multifractal generalizations of Hausdorff and packing measures.

Definition 7.4 Let $(X, \mathcal{F}, \mu)$ be a measure space and E, F in $\mathcal{F}$. We say that E is a subset of F μ-almost everywhere and write $E \subseteq F$ μ-a.e., if $\mu(F \setminus E) = 0$.

Consider the following sets

$$F = \Big\{x \in E;\ D^{\mathbf{q},t}_{\mu,\varphi}(x, E) = 1\Big\} \quad \text{and} \quad G = \Big\{x \in E;\ \Delta^{\mathbf{q},t}_{\mu,\varphi}(x, E) = 1\Big\}$$

Theorem 7.5 *Let φ be as in (7.1) and $\mu \in \mathcal{P}_D(\mathbb{R}^d)$. Let E be a $\mathcal{P}^{q,t}_{\mu,\varphi}$-measurable set with $\mathcal{P}^{q,t}_{\mu,\varphi}(E) < \infty$ and $\overline{\Delta}^{q,t}_{\mu,\varphi}(x, E) < +\infty$ for all $x \in E$. The following assertions are equivalent for any measurable subset B of E.*

1. $\mathcal{H}^{q,t}_{\mu,\varphi}(B) = \mathcal{P}^{q,t}_{\mu,\varphi}(B)$.
2. $\mathcal{P}^{q,t}_{\mu,\varphi}\Big(F \setminus B\Big) = 0$.
3. $\mathcal{P}^{q,t}_{\mu,\varphi}\Big(G \setminus B\Big) = 0$.

It is an easy consequence of the following lemmas.

Lemma 7.4 *Let φ be as in (7.1), $\mu \in \mathcal{P}_D(\mathbb{R}^d)$ and E be a Borel subset of* $\operatorname{supp}(\mu)$ *with $\mathcal{P}^{q,t}_{\mu,\varphi}(E) < \infty$. For all measurable subset B of E we have*

$$\mathcal{H}^{q,t}_{\mu,\varphi}(B) = \mathcal{P}^{q,t}_{\mu,\varphi}(B) \text{ if and only if } \mathcal{P}^{q,t}_{\mu,\varphi}\Big(G \setminus B\Big) = 0.$$

Proof. Without loss of the generality, we may assume that $\mathcal{P}^{\mathbf{q},t}_{\mu,\varphi}(B) > 0$. First, we suppose that $\mathcal{H}^{\mathbf{q},t}_{\mu,\varphi}(B) = \mathcal{P}^{\mathbf{q},t}_{\mu,\varphi}(B)$. By Theorem 7.3, we obtain $\Delta^{\mathbf{q},t}_{\mu,\varphi}(x, B) = 1$, for $\mathcal{P}^{\mathbf{q},t}_{\mu,\varphi}$-a.e. $x \in B$. By Lemmas 7.2 and 7.3, we obtain

$$\Delta^{\mathbf{q},t}_{\mu,\varphi}(x, B) = \Delta^{\mathbf{q},t}_{\mu,\varphi}(x, E), \text{ for } \mathcal{P}^{\mathbf{q},t}_{\mu,\varphi}\text{-a.e. } x \in B.$$

Now, assume that $\Delta^{\mathbf{q},t}_{\mu,\varphi}(x, E) = 1$, for $\mathcal{P}^{\mathbf{q},t}_{\mu,\varphi}$-a.e. $x \in B$. Then, we easily see that $\Delta^{\mathbf{q},t}_{\mu,\varphi}(x, E) = 1$, for $\mathcal{H}^{\mathbf{q},t}_{\mu,\varphi}$-a.e. $x \in B$. Using Lemma 7.2, we get

$$\Delta^{\mathbf{q},t}_{\mu,\varphi}(x, B) = \Delta^{\mathbf{q},t}_{\mu,\varphi}(x, E), \text{ for } \mathcal{H}^{\mathbf{q},t}_{\mu,\varphi}\text{-a.e. } x \in B.$$

We have $\Delta^{\mathbf{q},t}_{\mu,\varphi}(x, B) = 1$, for $\mathcal{H}^{\mathbf{q},t}_{\mu,\varphi}$-a.e. $x \in B$. From Lemma 7.3, we get $\Delta^{\mathbf{q},t}_{\mu,\varphi}(x, B) = 1$, for $\mathcal{P}^{\mathbf{q},t}_{\mu,\varphi}$-a.e. $x \in B$. Finally, Theorem 7.3 permits to get $\mathcal{H}^{\mathbf{q},t}_{\mu,\varphi}(B) = \mathcal{P}^{\mathbf{q},t}_{\mu,\varphi}(B)$.

Lemma 7.5 *Let φ be as in (7.1), $\mu \in \mathcal{P}_D(\mathbb{R}^d)$ and E be a $\mathcal{P}^{q,t}_{\mu,\varphi}$-measurable set with $\mathcal{P}^{q,t}_{\mu,\varphi}(E) < \infty$. Then*

$$\mathcal{P}^{q,t}_{\mu,\varphi}\Big(F \setminus G\Big) = 0.$$

Moreover, if $\overline{\Delta}^{q,t}_{\mu,\varphi}(x, E) < +\infty$ on E, we get

$$\mathcal{P}^{q,t}_{\mu,\varphi}\Big(G \setminus F\Big) = 0.$$

Proof. Without loss of generality, we may assume that $\mathcal{P}^{\mathbf{q},t}_{\mu,\varphi}(G) > 0$. By using Theorem 7.4, we have $\Delta^{\mathbf{q},t}_{\mu,\varphi}(x, G) = 1$, for $\mathcal{P}^{\mathbf{q},t}_{\mu,\varphi}$-a.e. $x \in G$. From Theorem 7.3, we obtain $D^{\mathbf{q},t}_{\mu,\varphi}(x, G) = 1$, for $\mathcal{P}^{\mathbf{q},t}_{\mu,\varphi}$-a.e. $x \in G$. Hence, $D^{\mathbf{q},t}_{\mu,\varphi}(x, G) = 1$, for $\mathcal{H}^{\mathbf{q},t}_{\mu,\varphi}$-a.e. $x \in G$. Using Lemma 7.2, we get

$$\overline{D}^{q,t}_{\mu,\varphi}(x,G) = \overline{D}^{q,t}_{\mu,\varphi}(x,E) \text{ and } \underline{D}^{q,t}_{\mu,\varphi}(x,G) = \underline{D}^{q,t}_{\mu,\varphi}(x,E), \text{ for } \mathcal{H}^{q,t}_{\mu,\varphi}\text{-a.e. } x \in G.$$

Then $D^{q,t}_{\mu,\varphi}(x,E) = 1$, for $\mathcal{H}^{q,t}_{\mu,\varphi}$-a.e. $x \in G$. By Lemma 7.3, $D^{q,t}_{\mu,\varphi}(x,E) = 1$, for $\mathcal{P}^{q,t}_{\mu,\varphi}$-a.e. $x \in G$. The remaining part may be proved by similar techniques.

7.5 Some Motivations, Applications, Examples and Discussions

In the present section, we propose to discuss and develop some motivating ideas and applications related to the present context. Some concepts are of theoretical aspects, however, some others will be of a purely applied aspect.

From a theoretical point of view, it is firstly worth noticing that, the theoretical results investigated in the present chapter remain extensible on general separable metric space where the Besicovitch covering theorem holds. Moreover, and already from the theoretical point of view, the density results obtained here may be applied as general forms of [118].

From a practical point of view, we recall that one of the most used sets and measures in physical applications are the so-called Moran sets and measures. The present chapter may be combined with [3] and [110] to show that the φ-mixed multifractal Hausdorff and packing measures are equivalent to Moran and cookie-cutter-like sets satisfying the strong separation condition for some prescribed measures μ and function φ.

Notice that our main results are adapted with the vectorial multifractal formalism introduced by Peyrière in [99]. A valuation on the metric space $(\mathbb{X},\rho)$ is a real function ξ defined on the set of balls of $\mathbb{X}$, subject to the condition that it goes to $+\infty$ as the radius goes to 0. In [99], Peyrière defined a more general multifractal analysis by considering quantities of type

$$\sum_i e^{-\left(\left\langle q,\chi(x_i,r_i)\right\rangle + t\xi(x_i,r_i)\right)}$$

where $\chi : \mathbb{X} \times [0,+\infty) \mapsto \mathbb{E}'$ is a function, $\mathbb{E}'$ is the dual of a separable real Banach space $\mathbb{E}$, $\langle .,. \rangle$ is the duality bracket between $\mathbb{E}$ and $\mathbb{E}'$ and $(\mathbb{X},\rho)$ is a metric space where the Besicovitch covering theorem holds. It is not difficult to verify that the all above main theorems hold for this vectorial multifractal analysis for some prescribed functions χ and ξ.

Let $(\mathbb{X},\rho)$ be a separable metric space, $\mathcal{B}$ stand for the set of balls of $\mathbb{X}$, and $\mathcal{F}$ for the set of maps from $\mathcal{B}$ to $[0,+\infty)$. The set of $\mu \in \mathcal{F}$ such that $\mu(B) = 0$ implies $\mu(B') = 0$ for all $B' \subseteq B$ will be denoted by $\mathcal{F}^*$. For such a μ, one defines its support $\text{supp}(\mu)$ to be the complement of the set $\bigcup\{B \in \mathcal{B} \mid \mu(B) = 0\}$. If moreover, we assume $(\mathbb{X},\rho)$ having the Besicovitch property, then the phenomena are the same for any functions μ in $\mathcal{F}^*$.

It is easy to check that we get the same values for fractal and multifractal measures and dimensions if we use just open balls or just closed balls and would not change the results. Note that the use of just closed balls is necessary where the use of open balls was erroneous when applying Vitali's theorems and in this case, we need to apply the Besicovitch covering theorem.

In the next parts, more motivations and examples will be discussed related to these conepts.

7.5.1 Why the Mixed Multifractal Measures, Densities, and their Regularities

Theoretically, the results here show that the computation of the fractal dimension is always possible when estimating well the mixed multifractal densities. Mixed multifractal analysis may also be applied to quantify the irregular dependence in the correlation between functions, time series, distributions, and measures. It is applied in many fields such as smart grids, smart cities, nano-materials, patterns, bioimages, ...etc.

Mixed multifractal analysis has been applied to understanding the joint movements for asset markets, long memory with mixture distributions. Dai in [35] served from mixed multifractal analysis of measures in the extraction of stock markets indices properties. Detrended fluctuation analysis has been also extended to mixed detrended fluctuation analysis by utilising the mixed multifractal analysis, and next applied for modeling the dependence between markets' indices, see for example [19], [25], [26], [28], [29], [37], [43], [46], [58], [59], [68], [70], [74], [75], [76], [85], [96], [113], [116], [131].

Mixed multifractal analysis appeared also in physics and statistics in different forms, but not really, strongly linked to the mathematical theory, such as in clustering topics where each attribute may be described by more than one type of measure. This leads researchers to apply measures well adopted for mixed-type data. See for example [52], [83].

Mixed multifractal concepts have been also extended to image processing as in [123], where texture characterization of natural images has been investigated showing that image intensity may fluctuate irregularly. In agricultural markets also, the authors in [120] applied the mixed densities to understand volatility and cross-correlation.

In the environment, there has been some interest in applying mixed multifractal analysis concepts to understand environmental factors such as [20] where the dynamics of atmospheric carbon emissions and industrial production index have been investigated by utilising joint multifractal analysis. Already in the study of environmental indices, and/or climate factors, the authors in [31] investigated the thermal structure of the mixed layer in the sea surface.

An application of mixed multifractal concepts has been conducted on biomedical images in [27]. See also [121]. Moreover, many practical studies have been developed by means of mixed densities such as climate factors, crashes propagation as in [1], [30], [50], [53], [98], [106]. These applications have been tackled by using the single multifractal analysis, and densities.

Mixed multifractal densities are also met in number theory and stochastic calculus such as [67] where a mixed analysis of Markov maps has been conducted with at a most countable number of branches. The authors in [115] served of mixed multifractal densities to understand multivariate time series with time-varying joint distribution characterized by the presence of self-affine structure. Besides, Combrexelle in [34] showed a potential benefit of mixed multifractal analysis in remote sensing applications. In [66] a mixed multifractal analysis of measures has been applied for mixed logical dynamical models in network traffic for the control of hybrid systems, such as multi-server ones.

In [41], [109], the authors revisited the Besicovitch covering theorem in the framework of mixed multifractal analysis, and established some results on multifractal densities in Tricot's sense and the regularities of measures.

7.5.2 Some Motivating Examples

Example 1. Graf and collaborators have widely studied the so-called h-Hausdorf measures and dimensions (see for example [62]). These measures are written on the form (for the Hausdorff measure for example),

$$\mathcal{H}^h(E) = \liminf_{\varepsilon \searrow 0} \left\{ \sum_i h(B(x_i, r_i));\ \ x_i \in E \subset \bigcup_i B(x_i, r_i),\ \ r_i < \varepsilon \right\},$$

where $h(r) = r^\alpha (\log|\log r|)^\beta$, with some constants α, β.

This variant of multifractal measures may be obtained from ours by just setting μ as the Lebesgue measure, and $\varphi(r) = \log(\log|\log r|)$.

For suitable choices of the parameters α, β, the authors in [63], succeeded in computing the exact Hausdorff dimension of some multifractal sets K_ω known as a random self-similar set depending on some set parameters w chosen randomly in some space Ω. They proved precisely that $0 < \mathcal{H}^h(K_\omega) < \infty$. Besides, in [5] raised an open problem about the gauge function φ to get a finite, infinite or zero value of $\mathcal{H}^{q,t}_{\mu,\varphi}(K)$. The present framework gives in fact an answer to this question. In [6] the author provided one case for the infinite value and one case for the zero one.

More investigations have been also conducted in [8], [47], [61], [62], [80], [122].

Example 2. A fascinating example has been provided in [82] about an eventual link between multifractal analysis of measures and functions to complex analysis, and especially the computation of the so-called long number of functions, differential forms, and currents.

For a bounded domain $\Omega \subset \mathbb{R}^n$ or $\mathbb{C}^n$, a function $F : \Omega \longrightarrow \mathbb{C}$, and $z \in \Omega$, the mean value of F around z is

$$\lambda_F(z, r) = \frac{1}{\sigma_n} \int_{S(0,1)} u(z + r\omega)\, d\sigma(\omega),$$

where σ_n is the area of the unit sphere in $\mathbb{R}^n$ (or $\mathbb{C}^n$) and $d\sigma$ is the Lebesgue measure on the sphere $\mathbb{S}^{n-1}$ in $\mathbb{R}^n$ (or $\mathbb{C}^n$). The Lelong number of F at the point z is defined by

$$\nu_F(z) = \lim_{r \to 0} \frac{\lambda_F(z, r)}{\log r}.$$

In some special cases, such as PSH functions, the computation of the long number is somehow easy. However, $\nu_F(z)$ remains unknown in general. The question is for what points z is possible to compute? And in the other cases, could it be possible to replace $\log r$ with a suitable function $\varphi(r)$? See [38], [39], [114] for more discussions.

Example 3. The last example deals with a general form of Weibull or Pareto distribution, which is already extended in statistics to mixed cases. Consider the density-like function

$$u(t) = -\frac{\gamma(\log t)^{\gamma-1}}{t} e^{-(|\log t|)^\delta} \chi_{]0,1[}(t)$$

where $\gamma > 0$ is a constant. Consider next the density measure $\mu(t) = u(t)dt$. It is easy to see that $\alpha_\mu(0) = -\infty$, and

$$\log \mu(B(x, r)) = -(\log x)^2 - \frac{2\log x}{x} r + r\varepsilon(r), \quad r \to 0, \ \ x \in]0, 1).$$

We immediately conclude that for all $\alpha > 0$,

$$\log \mu(B(x, r)) \sim (\log r^\alpha) \ \text{ as } r \to 0.$$

By setting $\varphi(r) = \log r(|\log r|)^{\gamma-1}$ and considering the measure

$$\nu_{q,t}(B(x,r)) = (\mu(B(x,r)))^q e^{t\varphi(r)},$$

we get

$$\frac{\nu_{q,t}(B(x,2r))}{\nu_{q,t}(B(x,r))} \sim \left[\frac{\mu(B(x,2r))}{\mu(B(x,r))}\right]^q e^{\gamma(\log 2)|\log r|^{\gamma-1}}, \quad \text{as } r \to 0.$$

If $\mu \in \mathcal{P}_D(\mathbb{R}^d)$ (a doubling measure, which is the case in all the original works in single multifractal analysis), we get

$$\frac{\nu_{q,t}(B(x,2r))}{\nu_{q,t}(B(x,r))} \sim C_{q,\mu} e^{\gamma(\log 2)|\log r|^{\gamma-1}} \longrightarrow +\infty \text{ (or 0) as } r \to 0.$$

This means that the measure $\nu_{q,t}$ which plays the role of the Gibbs measure supported on the singularity set of μ is no longer Gibbs in the present case. The new context permits the investigation of such types of measures on fractal sets for example.

7.5.3 Exact Dimensions of Some Non Gibbs Measures

To finish with this part on motivations and examples, we propose in this subsection to develop an example of exact computation of the fractal dimension of modified Cantor set (known as Moran-like set) by applying the present framework of mixed multifractal analysis, and where the classical one did not work. The example constructed here is a compilation of [82] which provided already an answer to a question raised in [5] on the gauge function φ leading to a finite, infinite, or zero value of $\mathcal{H}^{q,t}_{\mu,\varphi}(K)$ for the singularities set K?

The construction provided in [82], and recalled in a modified form here is in fact a combination of [21], [22], [23], [24], [61], [63], [62] on the construction of slef-similar type sets, and [14], [15] on the non-Gibbs-like measures. Let $a = (a_n^m)_{n,m}$ be a (suitable) sequence of non-negative real numbers. We start as in the construction of any Cantor-type set by an initial whole interval I, for example, $I = I_0 = [0, 1]$, affected with an initial total measure $\mu_0^0 = 1$. Next, the first iteration consists in cutting off an open interval of length a_1^1, and which is far from the extremities. We thus get two disjoint intervals I_1 and I_2, which will be affected by the masses μ_1^1 and μ_2^1 respectively. Next, we cut off from I_1 an interval of length a_1^2 and from I_2 an interval of length a_2^2. We get thus four intervals I_{11}, I_{12}, I_{21} and I_{22}, which will be affected respectively by the weighs $\mu_1^1\mu_1^2$, $\mu_1^1\mu_2^2$, $\mu_2^1\mu_1^2$ and $\mu_2^1\mu_2^2$ and so on. For a step $k \in \mathbb{N}$, we get 2^k intervals $I_{i_1 i_2 \ldots i_k}$, $(i_1 i_2 \ldots i_k) \in \{1, 2\}^k$ affected with a mass $\mu_{i_1 i_2 \ldots i_k} = \mu_{i_1}^1 \mu_{i_2}^2 \ldots \mu_{i_k}^k$, which leads thus to a probability measure μ supported by the Cantor set.

It is straightforward that μ is not a Gibbs measure (see for instance [82]). To compute the fractal dimension of its singularities set, we will apply the framework of the φ-mixed multifractal analysis. Assume for example that all the cells $I_{i_1,i_2,\ldots,i_n}$ have the same length

$$\rho_n = |I_{i_1,i_2,\ldots,i_n}| = e^{-e^{n\theta}},$$

for some $\theta \in \mathbb{R}$ fixed. The Cantor set obtained is no longer equivalent to any uniform construction known previously. Consider next the gauge function

$$\varphi(r) = -\log(1 + \log^+(\frac{1}{r})).$$

Assume finally that the masses μ_i^n satisfy

$$\mu_1^n = \delta \pm \varepsilon_n \quad \text{and} \quad \mu_2^n = \gamma \mp \varepsilon_n, \tag{7.17}$$

for some constants $\delta, \gamma \in]0,1[$ with $\delta \pm \gamma = 1$ and $(\varepsilon_n)_n \subset (0,1)$ such that $\delta \pm \varepsilon_n$ and $\gamma \pm \varepsilon_n$ lie in $]0,1[$. Technical calculus in [82] yield that

$$b_{\mu,\varphi}(p) = B_{\mu,\varphi}(p) = \Lambda_{\mu,\varphi}(p) = \frac{\log(\delta^p + \gamma^p)}{\theta}.$$

Denote next,

$$\alpha = \theta \frac{\delta^q \log \delta + \gamma^q \log \gamma}{\delta^q + \gamma^q}.$$

We get

$$\dim(X_{\mu,\varphi}(\alpha)) = \inf_p \Big(-\theta \frac{\delta^q \log \delta + \gamma^q \log \gamma}{\delta^q + \gamma^q} p + \frac{\log(\delta^p + \gamma^q)}{\theta} \Big).$$

This example joins also the results developed in [11], [12], [14], [15], where the computation of the fractal dimension of the Cantor type-set has been obtained by applying wavelet multifractal formalism for functions. The present framework permitted a simple way to do the job without the use of extra tools such as wavelets. Finally, it is worth it to note that in the case of constant sequences $(\mu_j^n)_n$, we join the results of [21], [22], [23], [24], [54], [55], [64], [65].

Denote

$$u(p) = \delta^p + \gamma^p, \quad H(p) = \frac{u'(p)}{u(p)}, \quad F_q(p) = \frac{1}{\beta} \log u(p) - \beta p H(q)$$

and for $s > 0$, $g_s(p) = \frac{1}{\beta} \log s - \beta H(p)$. We get

$$\dim \left(X_{\mu,\varphi}(\alpha) \right) = \inf_p \left(\frac{1}{\beta} \log u(p) - \beta p H(q) \right) = \inf_p F_q(p)$$

It remains now to evaluate this minimum. We have

$$F_q'(p) = \frac{1}{\beta} H(p) - \beta H(q) \text{ and } F_q''(p) = \frac{(\delta\gamma)^p}{\beta} \frac{(\log \delta - \log \gamma)^2}{(\delta^p + \gamma^p)^2}.$$

7.6 Conclusion

This chapter is concerned with some density estimations of vector-valued measures in the framework of the so-called mixed multifractal analysis. Precisely it is considered some Borel probability measures that are no longer Gibbs and introduced some mixed multifractal generalizations of densities in a framework of relative mixed multifractal analysis. Also, some results on multifractal regularities are developed in the new framework.

7.7 Appendix

Theorem 7.6 (Besicovitch covering theorem)

[17], [77]. There exists $\xi \in \mathbb{N}$ such that, for any subset A of $\mathbb{R}^d$ and any set of real numbers $(r_x)_{x\in A}$ satisfying

1. $r_x > 0, \forall\, x \in A$,
2. $\sup\limits_{x\in A} r_x < \infty$,

there exists ξ countable or finite subfamilies $B_1, \ldots, B_\xi$ of $\{B(x, r_x),\ x \in A\}$, such that

1. $A \subset \bigcup_i \bigcup_{B\in B_i} B$.
2. *B_i is composed of disjoint sets.*

Lemma 7.6 (Vitali's lemma)

[77]. Let X be a bounded compact metric space and $\mathcal{B}$ a set of closed balls in X, such that

$$\sup\left\{\dim(B),\ B \in \mathcal{B}\right\} < \infty.$$

Then, there exists a finite or countable sequence of disjoint balls $(B_i)_i \subset \mathcal{B}$, such that

$$\bigcup_{B\in\mathcal{B}} B \subset \bigcup_i (5B_i).$$

Theorem 7.7 (Vitali 1)

[77]. Let μ be a Radon measure on $\mathbb{R}^d$, $A \subset \mathbb{R}^d$ and $\mathcal{B}$ a family of closed balls, such that each point of A is the center of an arbitrarily small ball of $\mathcal{B}$, i.e.,

$$\inf\left\{r,\ B(x,r) \in \mathcal{B}\right\} = 0,\ \textit{for}\ x \in A.$$

Then, there exists a family of disjoint balls $(B_i)_i \subset \mathcal{B}$, such that

$$\mu\Big(A \setminus \bigcup_i B_i\Big) = 0.$$

Theorem 7.8 (Vitali 2)

[45] Let X be a metric space, E a subset of X and $\mathcal{B}$ a family of fine cover of E. Then, there exists either

1. *an infinite (centered closed ball) packing $\{B(x_i, r_i)\}_i \subset \mathcal{B}$, with $\inf\{r_i\} > 0$,*
   or
2. *a countable (possibly finite) centered closed ball packing $\{B(x_i, r_i)\}_i \subset \mathcal{B}$, such that for all $k \in \mathbb{N}$,*

$$E \setminus \bigcup_{i=1}^{k} B(x_i, r_i) \subset \bigcup_{i\geq k} B(x_i, 5r_i).$$

Figure 7.2 represents the graph of the function g_δ. It has a unique zero q_0. Furthermore, $q_0 \in (5, 5.5)$. Using the well known Newton-Raphson method, q_0 is estimated by $q_0 = 5.2266$.

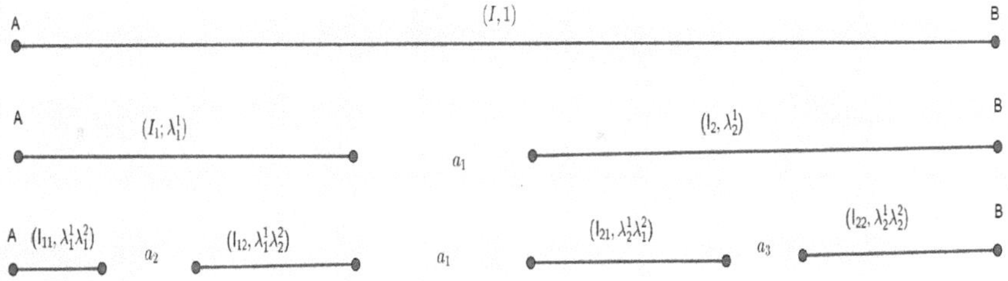

Fig. 7.1 The Cantor-type construction.

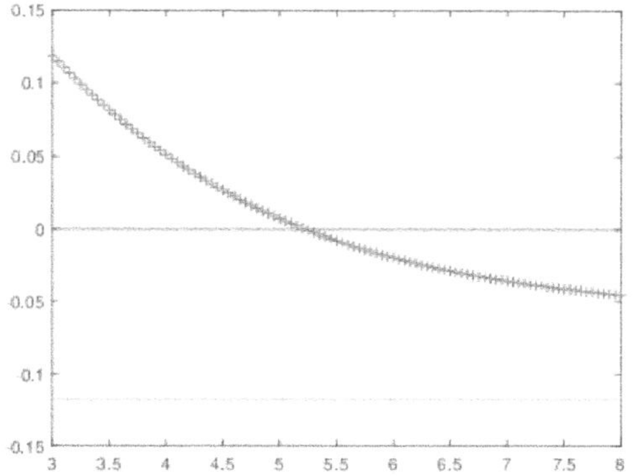

Fig. 7.2 Graph of g_δ.

Next, choosing $q = q_0+1$, the function F'_q has a unique minimum on $\mathbb{R}$ estimated also by Newton-Raphson method as $p_0 = -4.2624$. This is also illustrated by Figure 7.3.

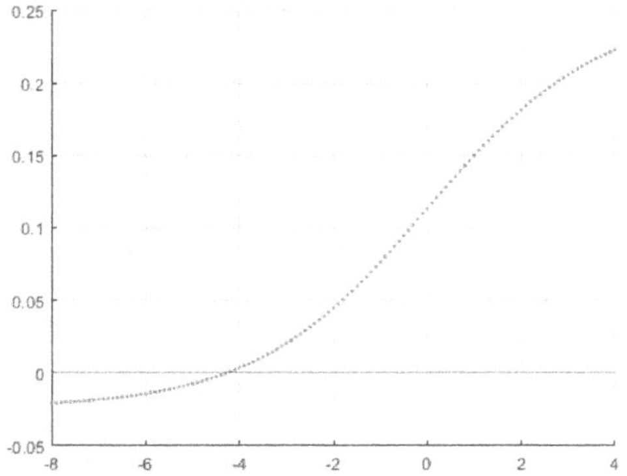

Fig. 7.3 Graph of F'_{q_0+1}.

Consequently, we get $\alpha = -0.3225$ and

$$dim(X_{\mu,\varphi}(0.3225)) = 0.1487.$$

Figure 7.4 confirms this value.

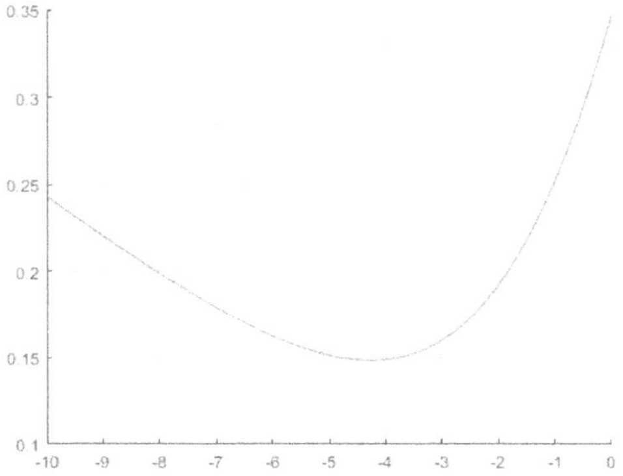

Fig. 7.4 Graph of F_{q_0+1}.

References

[1] Abry, P., H. Wendt and G. Didier. 2018. Detecting, and estimating multivariate self-similar sources in high-dimensional noisy mixtures. IEEE Workshop on Statistical Signal Processing (SSP 2018), Germany, 688–692. hal-02279354.

[2] Attia, N. and B. Selmi. 2019. Regularities of multifractal Hewitt-Stromberg measures. Commun. Korean Math. Soc., 34: 213–230.

[3] Attia, N., B. Selmi and Ch. Souissi. 2017. Some density results of relative multifractal analysis. Chaos, Solitons and Fractals, 103: 1–11.

[4] Aversa, V. and C. Bandt. 1990. The multifractal spectrum of discrete measures. Acta Univ. Carolinae-Math. Et Phys., 31: 5–8.

[5] Bandt, C., S. Graf and M. Zahle. 1995. Fractal Geometry and Stochastics. Birkhäuser Verlag.

[6] Bandt, C., S. Graf and M. Zahle. 2000. Fractal Geometry and Stochastics II. Birkhäuser Verlag.

[7] Barlow, M.T. and S.J. Taylor. 1992. Defining fractal subsets of $\mathbb{Z}^d$. Proc. London Math. Soc., 64: 125–152.

[8] Barral, J. and S. Seuret. 2015. Recent Developments in Fractals and Related Fields. Conference on Fractals and Related Fields III, île de Porquerolles, France.

[9] Barreira, L. 2001. Variational properties of multifractal spectra. Nonlinearity, 14: 259–274.

[10] Barreira, L. and B. Saussol. 2001. Variational principles and mixed multifractal spectra. Trans. AMS., 353: 3919–3944.

[11] Ben Mabrouk, A. 2005. Multifractal analysis of some non isotropic quasi-self-similar functions. Far East J. Dynamical Systems, 7: 23–63.

[12] Ben Mabrouk, A. 2008. On some nonlinear non isotropic quasi-self-similar functions. Nonlinear Dynamics, 51: 379–398.

[13] Ben Mabrouk, A. 2008. A higher order multifractal formalism. Stat. Prob. Lett., 78: 1412–1421.
[14] Ben Mabrouk, A., M. Ben Slimane and J. Aouidi. 2014. A wavelet multifractal fromalism for simultaneous singularities of functions. International Journal of Wavelets, Multiresolution and Information Processing, 12: 14.
[15] Ben Mabrouk, A., Mourad Ben Slimane and Jamil Aouidi. 2016. Mixed multifractal analysis for functions: General upper bound and optimal results for vectors of self-similar or quasi-self-similar of functions and their superpositions. Fractals, 24: 12.
[16] Ben Nasr, F. and I. Bhouri. 1997. Spectre multifractal de mesures boréliènnes sur $\mathbb{R}^d$. C. R. Acad. Sci. Paris, 325: 253–256.
[17] Besicovitch, A.S. 1938. On the fundamental geometrical properties of linearly measurable plane sets of points II. Math. Ann., 155: 296–329.
[18] Billingsley, P. 1965. Ergodic Theory and Information, John Wiley & Sons, Inc., New York, London, Sydney.
[19] Biswas, A., H.P. Cresswell and C.S. Bing. 2012. Application of multifractal, and joint multifractal analysis in examining soil spatial variation: A review. pp. 109–138. Chapter 6 *In*: Sid-Ali Ouadfeul (ed.). Fractal Analysis, and Chaos in Geosciences, InTechOpen, 109–138. DOI.org/10.5772/51437.
[20] Bozkus, S.K., H. Kahyaoglu and A.M.M. Lawali. 2020. Multifractal analysis of atmospheric carbon emissions, and OECD industrial production index. International Journal of Climate Change Strategies, and Management, 12: 411–430. DOI: 10.1108/IJCCSM-08-2019-0050.
[21] Cabrelli, C.A., F. Mendevil, U.M. Molter and R. Shonkwiler. 2004. On the Hausdorff h-measure of Cantor sets. Pacific Journal of Mathematics, 217: 45–59.
[22] Cabrelli, C.A., U.M. Molter, V. Paulauskas and R. Shonkwiler. 2004/05. The Hausdorff dimension of p-Cantor sets. Real Analysis Exchange, 30: 413–433.
[23] Cabrelli, C.A., K.E. Hare and U.M. Molter. 2002. Sums of Cantor sets yielding an interval. Journal of the Australian Mathematical Society. Series A, 73: 405–418.
[24] Cabrelli, C.A., K.E. Hare and U.M. Molter. 1997. Sums of Cantor Sets, Ergodic Theory and Dynamical Systems, 17: 1299–1313.
[25] Calvet, L. and A. Fisher. 2008. Multifractal volatility, theory, forecasting, and pricing. Academic Press Advanced Finance Series, 1st Ed.
[26] Cao, G., L.-Y. He and J. Cao. 2018. Multifractal Detrended Analysis Method, and its Application in Financial Markets, Springer.
[27] Castiglioni, P. and A. Faini. 2019. A fast DFA algorithm for multifractal multiscale analysis of physiological time series. Front. Physiol., 10: 18, Article: 115, DOI: 10.3389/fphys.2019.00115.
[28] Cattani, C. and J. Rushchitsky. 2007. Wavelet, and Wave Analysis as Applied to Materials with Micro or Nanostructure, World Scientific.
[29] Chandrasekhar, E., V.P. Dimri and V.M. Gadre. 2014. Wavelets, and Fractals in Earth System Sciences, Taylor & Francis Group.
[30] Chen, G. and Q. Cheng. 2017. Fractal density modeling of crustal heterogeneity from the KTB deep hole. J. Geophys. Res. Solid Earth, 122: 1919–1933.
[31] Chu, P.C. 1999. Multifractal analysis of the southwestern iceland sea surface mixed layer thermal structure. Thirteenth Symposium on Boundary Layers, and Turbulence, American Meteorological Society, 476–479. http://hdl.handle.net/10945/36212.
[32] Cole, J. 2000. Relative multifractal analysis. Choas Solitons Fractals, 11: 2233–2250.
[33] Cole, J. and L. Olsen. 2003. Multifractal variation measures and multifractal density theorems. Real Anal. Exch., 28: 501–514.

[34] Combrexelle, S. 2016. Multifractal analysis for multivariate data with application to remote sensing. Thèse de doctorat de l'université de Toulouse, Spécialité: signal, image, acoustique et optimisation.
[35] Dai, M., J. Hou, J. Gao, W. Su, L. Xi and D. Ye. 2016. Mixed multifractal analysis of China, and US stock index series. Chaos Solitons and Fractals, 87: 268–275.
[36] Das, M. 2005. Hausdorff measures, dimensions and mutual singularity. Trans. Am. Math. Soc., 357: 4249–4268.
[37] Dauphiné, A. 2012. Fractal Geography, John Wiley & Sons, Inc.
[38] Demailly, J.-P. 2012. Complex Analytic and Differential Geometry. Université de Grenoble I, Institut Fourier, France.
[39] Dolbeault, P., A. Iordan, G. Henkin, H. Skoda and J.-M. Trepreau. 2000. Complex analysis and geometry. International Conference in Honor of Pierre Lelong, Springer Basel AG.
[40] Douzi, Z. and B. Selmi. 2016. Multifractal variation for projections of measures. Chaos Solitons Fractals, 91: 414–420.
[41] Douzi, Z. and B. Selmi. 2019. Regularities of general Hausdorff and packing functions. Chaos, Solitons and Fractals, 123: 240–243.
[42] Douzi, Z. and B. Selmi. 2021. A relative multifractal analysis: Box-dimensions, densities, and projections, Quaestiones Mathematicae.
[43] Drozdz, S., R. Kowalski, P. Oswiecimk, R. Rak and R. Gebarowski. 2018. Dynamical variety of shapes in financial multifractality. Complexity, 7015721–13. DOI: 10.1155/2018/7015721.
[44] Edgar, G.A. 1994. Packing measures a gauge variation. Proceedings of the American Mathematical Society, 122: 167–174.
[45] Edgar, G.A. 2007. Centered densities and fractal measures. New York J. Math., 13: 33–87.
[46] Fan, Q., S. Liu and K. Wang. 2019. Multiscale multifractal detrended fluctuation analysis of multivariate time series. Physica A: Statistical Mechanics and its Applications, 532: 121864, DOI: 10.1016/j.physa.2019.121864.
[47] Fan, A.-H. and J. Schmeling. 2003. On fast Birkhoff averaging. Math. Proc. Cambridge Philos. Soc., 135: 443–467.
[48] Ben Mabrouk, A. and A. Farhat. 2021. A mixed multifractal analysis for quasi Ahlfors vector-valued measures. Fractals, Accepted 2021, ArXiv:2103.05466v1.
[49] Ben Mabrouk, A. and A. Farhat. 2021. Mixed multifractal densities for quasi-Ahlfors vector-valued measures. Fractals, Accepted 2021, ArXiv:2103.05246v1.
[50] de Figueiredo, B.C.L., G.R. Moreira, B. Stosic and T. Stosic. 2014. multifractal analysis of hourly wind speed records in petrolina, Northeast brazil. Rev. Bras. Biom. Sao Paulo., 32: 599–608.
[51] Edgar, G.A. 2007. Centered densities and fractal measures. New York J. Math., 13: 33–87.
[52] Gan, G., Ch. Ma and J. Wu. 2007. Data Clustering Theory, Algorithms, and Applications, ASA-SIAM Series on Statistics and Applied Probability, SIAM, Philadelphia, ASA, Alexandria, VA.
[53] Garcia-Marin, A.P., J. Estévez, F.J. Jiménez-Hornero and J.L. Ayuso-Munoz. 2013. Multifractal analysis of validated wind speed time series. Chaos: An Interdisciplinary Journal of Nonlinear Science, 23: 013133. DOI: 10.1063/1.4793781.
[54] Garcia, I. and L. Zuberman. 2012. Exact packing measure of central Cantor sets in the line. Journal of Mathematical Analysis and Applications, 386: 801–812.
[55] Garcia, I., U.M. Molter and R. Scotto. 2007. Dimension functions of Cantor sets. Proceedings of the American Mathematical Society, 135: 3151–3161.
[56] Gelfert, K. and D. Kwietniak. 2018. On density of ergodic measures and generic points. Ergod. Th. & Dynam. Sys., 38: 1745–1767.

[57] Gelfert, K. and M. Rams. 2009. The Lyapunov spectrum of some parabolic systems. Ergod. Th. & Dynam. Sys., 29: 919–940.

[58] Ghanbarian, B. and A.G. Hunt. 2017. Fractals Concepts, and Applications in Geosciences, Taylor, and Francis Group.

[59] Ghosh, D., S. Samanta and S. Chakraborty. 2019. Multifractals, and Chronic Diseases of the Central Nervous System, Springer.

[60] Glasscock, D. 2016. Marstrand-type theorems for the counting and mass dimensions in $\mathbb{Z}^d$. Combinatorics, Probability and Computing, 25: 700–743.

[61] Graf, S. 1991. Random fractals, Lectures presented at teh School on Measure Theory and Real Analysis, Gradp, Italy, October, 14–25, 64 pages.

[62] Graf, S. and H. Luschgy. 2000. Foundations of Quantization for Probability Distributions, Springer, Lecture Notes in Mathematics 1730.

[63] Mauldin, R.D., S. Graf and S.C. Williams. 1987. The exact Hausdorff dimension in random recursive constructions. Proc. Nati. Acad. Sci. USA, 84: 3959–3961.

[64] Hare, K., F. Mendivil and L. Zuberman. 2013. The sizes of rearrangements of Cantor sets. Canadian Mathematical Bulletin, 56: 354–365.

[65] Hare, K. and L. Zuberman. 2010. Classifying Cantor sets by their multifractal spectrum. Nonlinearity, 23: 2919–2933.

[66] Hong, L., Y. Tie and W. Lanlan. 2010. Network traffic prediction based on multifractal MLD model. International workshop on Chaos-fractal theory, and its applications. IEEE Computer Society, pp. 466–470. DOI:10.1109/IWCFTA.2010.109.

[67] Jaerisch, J. and H. Takahasi. 2020. Mixed multifractal spectra of Birkhoff averages for non-uniformly expanding one-dimensional Markov maps with countably many branches, arXiv:2004.04347v2.

[68] Janahmadov, A.K. and M.Y. Javadov. 2016. Synergetics, and Fractals in Tribology (Materials Forming, Machining, and Tribology), Springer.

[69] Jarvenpaa, E., M. Jarvenpaa, A. Kaenmaki, T. Rajala, S. Rogovin and V. Suomala. 2010. Packing dimension and Ahlfors regularity of Porus sets in metric spaces. Math. Z., 266: 83–105.

[70] Jiang, Z.-Q., W.-J. Xie, W.-X. Zhou and D. Sornette. 2019. Multifractal analysis of financial markets: A review. Reports on Progress in Physics, 82: ID 125901, 105 pages.

[71] Jinrong, L., Y. Zuguo and R. Fuyao. 2000. Measures and their dimension spectrums for Cookie-Cutter in $\mathbb{R}^d$. Acta Mathe. Appl. Sinica., 16: 10–21.

[72] Jordan, T. and M. Pollicott. 2007. Multifractal analysis and the variance of Gibbs measures. J. Lond. Math. Soc., 76: 57–72.

[73] Jordan, T. and M. Rams. 2011. Multifractal analysis of weak Gibbs measures for non-uniformly expanding C^1 maps. Ergod. Th. & Dynam. Sys., 31: 143–164.

[74] Liu, R. and T. Lux. 2015. Non-homogeneous volatility correlations in the bivariate multifractal model. The European Journal of Finance, 21: 971–991.

[75] Liu, R. and T. Lux. 2017. Generalized method of moment estimation of multivariate multifractal models. Economic Modelling, 67: 136–148.

[76] Manshour, P. 2020. Nonlinear correlations in multifractals: Visibility graphs of magnitude, and sign series. Chaos, 30: 013151–9.

[77] Mattila, P. 1995. The Geometry of Sets and Measures in Euclidean Spaces, Cambridge University Press, Cambrdige.

[78] Mattila, P. and R.D. Mauldin. 1997. Measure and dimension functions: measurablility and densities. Math. Proc. Camb. Phil. Soc., 121: 81–100.

[79] Mattila, P. and P. Saaramen. 2009. Ahlfors-David regular sets and bilipschitz maps. Annales Academiæ Scientiarum Fennicæ Mathematica, 34: 487–502.

[80] Mauldin, R.D., S. Graf and S.C. Williams. 1987. Exact Hausdorff dimension in random recursive constructions. Proc. Nati. Acad. Sci. USA, 84: 3959–3961.

[81] Menceur, M., A. Ben Mabrouk and K. Betina. 2016. The multifractal formalism for measures, review and extension to mixed cases. Anal. Theory Appl., 32: 77–106.

[82] Menceur, M. and A. Ben Mabrouk. 2019. A joint multifractal analysis of vector valued non Gibbs measures. Chaos, Solitons and Fractals, 126: 1–15.

[83] Meneveau, Ch., K.R. Sreenivasan, P. Kailasnath and M.S. Fan. 1990. Joint multifractal measures: Theory and applications to turbulence. Physical Review A, 41: 894–913.

[84] Morse, A.P. and J.F. Randolph. 1944. The ϕ-rectifiable subsets of the plane. Am. Math. Soc. Trans., 55: 236–305.

[85] Nakayama, T. and K. Yakubo. 2003. Fractal Concepts in Condensed Matter Physics, Springer.

[86] Olsen, L. 1994. Random geometrically graph directed self-similar multifractals. Pitman Research Notes in Mathematics Series 307, Longman Scientific & Technical.

[87] Olsen, L. 1995. A multifractal formalism. Adv. Math., 116: 82–196.

[88] Olsen, L. 1996. Multifractal dimensions of product measures. Math. Proc. Camb. Phil. Soc., 120: 709–734.

[89] Olsen, L. 1998. Self-affine multifractal Sierpinski sponges in $\mathbb{R}^d$. Pacific J. Math., 183: 143–199.

[90] Olsen, L. 2000. Dimension inequalities of multifractal Hausdorff measures and multifractal packing measures. Math. Scand., 86: 109–129.

[91] Olsen, L. 2000. Integral, probability, and fractal measures by G. Edgar, Springer, New York, 1998. Bull. Amer. Math. Soc., 37: 481–498.

[92] Olsen, L. 2005. Mixed generalized dimensions of self-similar measures. J. Math. Anal. and Appl., 306: 516–539.

[93] Olsen, L. 2008. Multifractal analysis of divergence points of deformed measure theoretical Birkhoff averages IV: Divergence points and packing dimension. Bull. Sci. Math., 132: 650–678.

[94] Olsen, L. and S. Winter. 2007. Multifractal analysis of divergence points of deformed measure theoretical Birkhoff averages. II: Non-linearity, divergence points and Banach space valued spectra. Bull. Sci. Math., 131: 518–558.

[95] O'Neil, T.C. 2000. The multifractal spectra of projected measures in Euclidean spaces. Chaos, Solitons and Fractals, 11: 901–921.

[96] Ouadfeul, S.-A. 2012. Fractal Analysis, and Chaos in Geosciences. InTech Open.

[97] Pajot, H. 1996. Sous-ensembles de courbes Ahlfors-règulières et nombres de Jones. Publicacions Matematiques, 40: 497–526.

[98] Pavon-ominguez, P., F.J. Jimenez-Hornero and E. Gutierrez-de-Rave. 2013. Multifractal analysis of ground-level ozone concentrations at urban, suburban, and rural background monitoring sites in South-western Iberian Peninsula. Atmospheric Pollution Research, 4: 229–237.

[99] Peyrière, J. 2004. A vectorial multifractal formalism. Proc. Sympos. Pure Math., 72: 217–230.

[100] Pesin, Y. and V. Climenhaga. 1982. Lectures on fractal geometry and dynamical systems. AMS Mathematics Adavanced Study, Students Mathematical Library, 52.

[101] Pesin, Y.B. 1997. Dimension Theory in Dynamical Systems, Contemporary Views and Applications, Chicago Lectures in Mathematics.

[102] Preiss, D. 1987. Geometry of measures in $\mathbb{R}^n$: Distribution, rectifiablity and densities. Ann. Math., 125: 537–643.

[103] Qu, Ch. 2012. The lower densities of symmetric perfect sets. Anal. Theory Appl., 28: 377–384.

[104] Raymond, X.S. and C. Tricot. 1988. Packing regularity of sets in n-space. Math. Proc. Camb. Philos. Soc., 103: 133–145.

[105] Rigot, S. 2001. Ensembles quasiminimaux pour le périmètre et rectifiabilité uniforme, Sémin. Équ. Dériv. Partielles, 2000–2001, Exp. No. IX, 13 pp., École Polytech., Palaiseau, 13 pages.

[106] Scheuring, I. and R.H. Riedi. 1994. Application of multifractals to the analysis of vegetation pattern. Journal of Vegetation Science, 5: 489–496.

[107] Selmi, B. 2016. A note on the effect of projections on both measures and the generalization of q-dimension capacity. Probl. Anal. Issues Anal., 5(23): 38–51.

[108] Selmi, B. 2018. Some results about the regularities of multifractal measures. Korean J. Math., 26: 271–283.

[109] Selmi, B. 2019. On the strong regularity with the multifractal measures in a probability space. Anal. Math. Phys., 9: 1525–1534.

[110] Selmi, B. 2021. The relative multifractal densities: A review and application. Journal of Interdisciplinary Mathematics, 24: 1627–1644.

[111] Selmi, B. and N. Yu. 2017. Svetova, on the projections of mutual $L^{q,t}$-spectrum. Probl. Anal. Issues Anal., 6(24): 94–108.

[112] Selmi, B. 2020. Multifractal dimensions of vector-valued non-Gibbs measures. Gen. Lett. Math., 8: 51–66.

[113] Seuront, L. 2010. Fractals, and Multifractals in Ecology, and Aquatic Science, Taylor, and Francis Group.

[114] Siu, Y.-T. 1973. Analycity of sets associated to Lelong numbers and the extension of meromorphic maps. Bulletin of the American Mathematical Society, 79: 1200–1205.

[115] Taylor, S.J. 1995. The fractal analysis of Borel measures in $\mathbb{R}^d$. J. Fourier. Anal. and Appl., Kahane Special Issue, 553–568.

[116] Tolotti, M. 2012/2013. Fractal, and multifractal models for price changes. An attempt to manage the Black Swan, Corso di Laurea magistrale in Economia e Finanza, Universita CaFoscari Venisia.

[117] Tricot, C. 1999. Courbes et Dimension Fractale, Springer, 2nd Edition.

[118] Tricot, C. 2010. General Hausdorff functions and the notion of one-sided measure and dimension. Ark. Math., 48: 149–176.

[119] Vojack, R. and J. Lévy Véhel. 1996. Higher order multifractal analysis. INRIA, Rapport de Recherches, 2796: 34.

[120] Wang, H.-Y. and Y.-S. Feng. 2020. Multivariate correlation analysis of agricultural futures, and spot markets based on multifractal statistical methods. J. Stat. Mech., 2020: 073403. DOI.org/10.1088/1742-5468/ab900f.

[121] Wang, J., W. Shao and J. Kim. 2020. Multifractal detrended cross-correlation analysis between respiratory diseases, and haze in South Korea. Chaos, Solitons, and Fractals, 135: 109781, 10 pages.

[122] Wang, X. and M. Dai. 2010. Mixed quantization dimension function and temperature function for conformal measures. International Journal of Nonlinear Science, 10: 24–31.

[123] Wendt, H., S. Combrexelle, Y. Altmann, J.-Y. Tourneret, S. McLaughlin and P. Abry. 2018. Multifractal analysis of multivariate images using Gamma Markov random field Priors. SIAM J. Imaging Sciences, 11: 1294–1316.

[124] Xu, M. and Sh. Wang. 2011. The boundedness of bilinear singular integral operators on Sierpinski Gaskets. Anal. Theory Appl., 27: 92–100.

[125] Xu, Sh. and W. Xu. 2012. Note on the paper an negative answer to a conjecture on the self-similar sets satisfying the open set condition. Anal. Theory Appl., 28: 49–57.

[126] Xu, Sh., W. Xu and D. Zhong. 2012. Some new iterated function systems consisting of generalized contractive mappings. Anal. Theory Appl., 28: 269–277.
[127] Xu, Sh., W. Xu and Z. Zhou. 2015. Some results on the upper convex densities of the self-similar sets at the contracting-similarity fixed points. Anal. Theory Appl., 28: 92–100.
[128] Ye, Y.-L. 2007. Self-similar vector-valued measures. Adv. Appl. Math., 38: 71–96.
[129] Yuan, Y. 2015. Spectral self-affine measures on the generalized three Sierpinski Gasket. Anal. Theory Appl., 31: 394–406.
[130] Zeng, Ch., D. Yuan and Sh. Xui. 2012. The Hausdorff measure of Sierpinski carpets basing on regular Pentagon. Anal. Theory Appl., 28: 27–37.
[131] Zhang, Z.-X. 2016. Rock Fracture, and Blasting Theory, and Applications, Elsevier.
[132] Zhou, Z. and L. Feng. 2011. A theoretical framework for the calculation of Hausdorff measure self-similar set satisfying OSC. Anal. Theory Appl., 27: 387–398.
[133] Zhu, Z. and Z. Zhou. 2014. A local property of Hausdorff centered measure of self-similar sets. Anal. Theory Appl., 30: 164–172.

Chapter 8

Multifractal Dimensions and Fractional Differentiation in Automated Edge Detection on Intuitionistic Fuzzy Enhanced Image

VP Ananthi,[1] C Thangaraj[2] and D Easwaramoorthy[2,*]

8.1 Introduction

The process of dividing an image into meaningful regions that are suitable for further processing is called segmentation. This is a fundamental part of image analysis which finds its application in medical image analysis, pattern recognition, object detection, computer vision, etc. There are numerous approaches to segment images which include thresholding based techniques, region based segmentation, edge detection, clustering based techniques. The simplest of all these techniques is the thresholding-based segmentation technique [32, 57, 58, 61, 63, 67, 69]. Detection of edges in an image is concerned with illumination of light during capturing as well as its geometrical properties and noise/disturbance.

Edges are important structures in an image which divides an image into regions and helps in understanding the various elements in an image. These are identified by a sudden change in the intensity values of pixels. Edge-detection is a low level image processing task which aims to find all the edge pixels in an image. Despite huge number of edge-detection techniques available in literature [64], edge detection still remains a major research area.

Generally, differential operators are used to detect edges. The two main challenges in edge-detection task, which most of these state-of-art techniques fail to surmount are, misinterpretation of noise pixels as edge pixels and the non-detection of thin/fainted edges. Various edge detection techniques have been proposed over the years, but the common approach is to apply the first or second derivative. Based on this, the edge detection can be classified as gradient edge detectors (first order derivative), Laplacian method (second order derivative), or Gaussian edge detectors [13]. In the fuzzy domain, there are edge detection methods that have been developed for grayscale images [15] and [14]. In Gonzalez et al. [12], an improved method for processing edge detection based on general type-2 fuzzy sets (GT2 FSs) applied to grayscale images is proposed. According to the results presented in [12], the author demonstrated that the edge detection approach based on GT2 FSs outperformed the results obtained by the methods based on interval type-2 fuzzy sets (IT2 FSs) [11], type-1 fuzzy sets (T1 FSs) [11], and of course the traditional edge detection methods.

The common process of edge detection includes two terms, namely edge strength and edge enhancement. Further, in the edge enhancement, local edges of the image will be highlighted using edge operator. On the other hand, the threshold should be optimum for true edge point extraction because the image may be noisy or blurred as well as the fact the edges may not be continued [10]. Some edge points are removed from the edge point set to fill with another obtained linking edge point set into lines [9].

[1] Department of Mathematics, Gobi Arts & Science College, Gobichettipalayam - 638 453, Erode, Tamil Nadu, India.
[2] Department of Mathematics, School of Advanced Sciences, Vellore Institute of Technology, Vellore 632 014, Tamil Nadu, India.
Emails: ananthi@gascgobi.ac.in; ctrmath1@gmail.com
* Corresponding author: easandk@gmail.com

Generally, image edges and boundaries are fuzzy or vague. The main disadvantages of direct edge detection methods are that the detected edges may be broken and discontinued. Further, a preprocessing operation is required to get clear and true edges from the image. Preprocessing is a very significant step that plays a vital role in various image processing areas. Image enhancement can be performed as preprocessing technique. The significant features or partially visible features will be highlighted and preprocessing will suppress the unnecessary information that are not related to the image processing task [9]. There are mainly two types of enhancements, viz., contrast enhancement and edge enhancement. Moreover, edge enhancement is mostly used to detect edge and improve the quality of edges and boundaries of the images, while contrast enhancement improves the quality of whole image.

A variety of techniques have been proposed for detecting edges. The major application areas of edge detection techniques are the process of image segmentation, object detection, image enhancement, image compression, robotics, military, medical diagnosis, meteorology, computer vision, pattern recognition, geography, etc. Chaira et al. [8] proposed a new measure using intuitionistic fuzzy sets (IFSs) and its application to edge detection, which is a new technique, and provide comparatively better results. A prominent characteristic of IFSs is that it appoints to every element membership as well as non-membership degree with certain amount of hesitation degree [7]. Implementation of IFS have made a significant improvement in edge detection, which has been suggested in [6].

In recent years, the subject of fractional calculus has been of interest owing to its vast physical applications. It has found inroads even into the domain of image processing. Recently, researchers are working on the edge detectors based on fractional differentiation which are found to detect edges better than the ordinary integer order differential operators [54, 60, 62, 63, 70]. A difficulty in using the fractional operator is in choosing the appropriate order of the fractional derivative. It has been found that the response of fractional differential operator to a frequency component is different for different orders of the fractional differential operator [60]. Hence, the edges obtained vary for different orders of fractional differential operators. Thus there is an uncertainty in the selection of suitable order for each image. Edge image generated using a fractional differential operator of uncertain order is also fuzzy in nature. Since fuzzy sets have the ability to handle uncertainty, the resulting uncertain edge image is fuzzified.

The characteristics of images vary due to various situations such as the ill-posed of the physics solution, the real time of its application and so on. Therefore, detecting edge in images with varying factors is difficult to attain instantly without practicing it practically [55]. Compared to the integral order calculus, the implementation of the fractional order calculus is more complex, but it extends the order of the differential operator and has more freedom degree and flexibility [4]. Furthermore, the fractional differentiation can enhance the edges, texture and preserve the smooth areas of the image. While processing the edges, textures and smooth areas of image, the fractional derivatives can construct the masks with the value in eight directions [5]. One of the difficulties in using fractional order derivative is in the selection of order for a particular image. Hence this chapter concentrates on the automated selection of order for the fractional derivative concerned to the image by using minimum intuitionistic fuzzy divergence (IFD) between the images by changing the order.

A fractal is a rough or fragmented geometric shapes of natural features as well as other complex objects that traditional Euclidean geometry fails to analyze. According to Mandelbrot idea, the objects like mountains, clouds and whatever complex related sets; the Euclidean geometry was not enough to analysis of these complex type objects. Therefore, the complex related objects and these kinds of phenomenon are only delineated with respect to the fractal geometry. Fractal geometry is a traditional approach to characterize the complex patterns found in nature by using the property of self-similarity, which was originally explored by Mandelbrot in 1975. The fractal was derived from the Latin word, 'fractus'means broken or fractured, and to describe objects that were too irregular or complex to fit into a traditional geometrical setting. Mandelbrot defined a fractal mathematically as a set with Hausdorff dimension strictly exceeds its topological dimension [71]–[77].

Fractal dimension analyzes the irregularity of the given object with the various scaling properties. The concept of fractal dimension can be applicable in the measurement and categorization of complex shapes and textures. There are numerous research works that have been described in medical image analysis employed with the fractal analysis. Fractal dimension is insufficient to wards characterizing the object having a more complex and inhomogeneous scaling properties. Therefore, the Generalized Fractal Dimensions (GFD) to analyze the performance of the edge detection in standard and representative images. GFD can be applied to compare the proposed method with the Sobel method, Prewitt method, Roberts method and Canny method for the various complex images to analyze the complexity of the edge detection process.

The structure of this chapter is framed as five sections including the Introduction in this Section 8.1. Section 8.2 introduces some basic ideas related to IFS, the fractional derivatives and the multifractal analysis. Steps of the proposed algorithm have been discussed in the Section 8.3. Experimental results are discussed in the Section 8.4. Section 8.5 draws concluding statements and some future works.

8.2 Preliminaries

Brief idea about the fuzzy set, the generation of IFS from it along with the fractional derivative and the multifractal dimensions analysis have been discussed below.

8.2.1 Intuitionistic Fuzzy Set

Image processing by IFSs mainly needs membership and non-membership function. Now let us see briefly about the construction by initiating from FSs. Consider a finite set $X = \{x_1, x_2, x_3, ..., x_n\}$. A fuzzy set [3] F of X may be mathematically written as

$$F = \{(x, \mu_F(x))|x \in X\},$$

where the function $\mu_F(x) : X \longrightarrow [0, 1]$ represents the membership degree of an element x in X. Therefore, the non-membership degree of x is $1 - \mu_A(x)$. Support of a fuzzy set F in X is defined as $Supp(F) = \{x \in X|\mu_F(x) > 0\}$. A fuzzy set is said to be a fuzzy singleton if its support is a single point. Atanassov [2] generalized fuzzy sets as IFS. An IFS F in X can be mathematically symbolized as

$$F = \{(x, \mu_F(x), \nu_F(x))|x \in X\},$$

where the functions $\mu_F(x), \nu_F(x) : X \longrightarrow [0, 1]$ represents the degree of membership and non-membership of an element x in X respectively, with the essential condition $0 \leq \mu_F(x) + \nu_F(x) \leq 1$.

On observation, it is clearly seen that FSs is a peculiar case of IFS. A new parameter $\pi_F(x)$, which originates due to lack of knowledge called hesitation degree, has been introduced by Szmidt and Kacpryzk [1] while computing the distance between FSs. IFS is defined as follows based on the hesitation degree

$$F = \{(x, \mu_F(x), \nu_F(x), \pi_F(x))|x \in X\}$$

where the condition $\mu_F(x) + \nu_F(x) + \pi_F(x) = 1$ holds.

8.2.2 Overview of Fractional Calculus

Fractional calculus is an area of mathematical physics dealing with integrals and derivatives of positive order. This broad area of research had its embryo in past 300 years while Leibniz answered a question raised by L'Hospital whether there exists a generalization of integer order derivative? [46]. Liouville, Riemann, Weyl innovation stepped as a major role in the growth of fractional calculus. In recent days, Xue [47] have evoked this field in biomedical engineering. Fractional calculus have been incorporated in control theory by Richard [48, 49, 51, 52]. Oustaloup [53] developed fractional calculus in the area of physics. In the field of mechanics also fractional calculus have been investigated by Chen [50]. This non-integer derivative have been analyzed after the death of Leibniz till Euler's research. Euler [39] introduced a more generalized fractional of internal numbers as a Gamma function. After Euler and Lagrange stepped into the area of fractional calculus by presenting the principle of exponents for derivatives of integer order [38].

$$\frac{d^p}{dy^p}\frac{d^q}{dy^q} = \frac{d^{p+q}}{dy^{p+q}}, p, q \in N$$

Lacroix [37] defined a fractional derivative of a function $f(x) = x^q,\ q \in N$ with the use of Gamma function, quoted as below:

$$\frac{d^p}{dy^p}f(y) = \frac{q!}{(q-p)!}y^{q-p}, \quad q \geq p. \tag{8.1}$$

The above equation (8.1) can be rewritten as

$$\frac{d^p}{dy^p}f(y) = \frac{\Gamma(q+1)}{\Gamma(q-p+1)}y^{q-p}, \quad q \geq p \tag{8.2}$$

For example, $f(y) = y, p = 1/2, q = 1$, then by using equation (8.2), we get

$$\begin{aligned}\frac{d^p}{dy^p}y &= \frac{\Gamma(2)}{\Gamma(1-1/2+1)}y^{1-1/2}\\ &= \frac{1}{\Gamma(3/2)}y^{1/2}\\ &= 2\sqrt{\frac{y}{\pi}}\end{aligned}$$

Similarly in 1822, Fourier [36] derived more obscure fractional derivative defined as

$$\frac{d^k}{dy^k}f(y) = \frac{1}{2\pi}\int_{-\infty}^{\infty} f(x)dx \int_{-\infty}^{\infty} q^i cos(q(y-x) + \frac{i\pi}{2})dq,$$

where k is a positive or negative number. This derivative is applicable for organized functions that may not be a power or integral function. Liouville [33–35] developed various definitions of fractional derivative and it was employed in the series as follows:

$$s(y) = \sum_{i=0}^{\infty} k_i e^{b_i y}$$

Classical integer order differentiation of e^{by} can be written as

$$\frac{d^n}{dy^n}e^{by} = D^n e^{by}.$$

This can be generalized as a fractional derivative by changing $n \in N$ with $p \in C$ as

$$D^p s(y) = \sum_{i=0}^{\infty} k_i\, b_i^p\, e^{b_i y} \tag{8.3}$$

The above equation (8.3) is limited to the values of p and it takes the series to converge. For instance, $f(y) = \frac{1}{y^\alpha}$, where α is an arbitrary variable, Liouville defined a fractional derivative of order p, as

$$D^p y^{-\alpha} = (-1)^p \frac{\Gamma(\alpha + p)}{\Gamma(\alpha)} y^{-\alpha-p}$$

Taylor series is generalized by Riemann as deduced as follows:

$$D^p f(y) = \frac{1}{\Gamma(p)} \int_k^y (y-t)^{p-1} f(t)dt + \phi(y) \tag{8.4}$$

The function $\phi(y)$ in the above equation (8.4) is provided to furnish the retarding confusion of the lower order limit k of the integration, which is related to the complementary function. If $k = 0$ and $\phi(y) = 0$, then the above equation (8.4) is called as Riemann-Liouville fractional integral [30, 31]. Letnikov's work in [29] is an extension of Sonin's [28], which have been defined by the use of Cauchy's integral formula. Integer order derivative is defined as an integral formula depicted below:

$$\frac{d^m}{dy^m} f(y) = \frac{m!}{2\pi i} \int \frac{f(s)}{(t-s)^{m+1}} ds$$

Fractional derivative is a generalization of the above equation (8.2.2) by replacing the factorial $\beta = \Gamma(\beta + 1)$ as a Gamma function. Laurent's work is extended and fractional ordered derivative have been emerged as [27]

$$_kD_y^{-q} f(y) = \frac{1}{\Gamma(q)} \int_k^y (y-s)^{q-1} f(s)ds \tag{8.5}$$

$$_kD_y^{q} f(y) = \frac{d^n}{dx^n}\left(\frac{1}{\Gamma(1-q)} \int_k^y (y-s)^{q} f(s)ds\right)$$

Another representation of fractional derivative Grunwald [26] and Letnikov [25] is

$$D_y^q f(y) = \lim_{h\to 0} \frac{(\Delta_h^q f)(y)}{h^p} \tag{8.6}$$

In the above equation (8.6),

$$(\Delta_h^q f)(y) = (-1)^i \sum_{i=0}^{\infty} \binom{p}{i} f(y - ih), p > 0,$$

and $\binom{p}{i}$ is a binomial coefficient.

8.2.3 Recent Scenario in Fractional Calculus

Weyl [24] utilized a periodic function $\psi(y)$ as a Fourier transformation depicted as follows:

$$\psi(y) = \sum_{s=-\infty}^{\infty} \psi_s e^{isy},$$

$$\psi_s = \frac{1}{2\pi} \int_{-\infty}^{\infty} e^{isy} \psi(y) dy,$$

and fractional integration is given by

$$I_{\pm}^{p} \psi(y) = \frac{1}{2\pi} \int_{-\infty}^{\infty} \Phi_{\pm}^{p}(y-s) \psi(s) ds. \tag{8.7}$$

For $0 < p < 1$, the above equation (8.7) is written as

$$I_{+}^{p} \psi(y) = \tfrac{1}{\Gamma(p)} \int_{-\infty}^{y} (y-1)^{p-1} \psi(s) ds$$

$$I_{-}^{p} \psi(y) = \tfrac{1}{\Gamma(p)} \int_{y}^{\infty} (y-1)^{p-1} \psi(s) ds$$

and the above set of equations are convergent in $[-\infty, \infty]$. Weyl fractional integral is a particular case of Riemann-Liouville fractional integral equation (8.5), when $k = -\infty$. Grunwald-Letnikov fractional derivative have been represented as integral by Marchaud [23] as

$$D_y^q f(y) = K \int_0^{\infty} \frac{(\Delta_s^l f)(s)}{s^{1+q}} ds, q > 0,$$

where $(\Delta_s^l f)(s)$ is a finite difference of order $l > q$ and constant K. Wantanabe [22] represented fractional derivative for analytic functions u and v as

$$D^q(uv) = \sum_{i=-\infty}^{\infty} \binom{q}{i+p} D^{q-p-i} u \, D^{p+i} v, p \in R.$$

Fractional derivative is defined by Caputo in [21] for an $(n-1)$ continuously differentiable function u as

$$D^q u(y) = \frac{1}{\Gamma(n-q)} \int_0^x (x-t)^{n-q-1} \left(\frac{d}{dt}\right)^n u(t) dt,$$

with zero initial condition.

8.2.4 Edge Detection Based on Fractional Derivative

Integer order derivative of order 1 and 2 are the frequently used as classical edge detectors.

In past decades, fractional calculus have attained its greater height in the areas of applied fields such as dynamical systems and artificial intelligence (AI). For example, in lower level AI techniques such as image

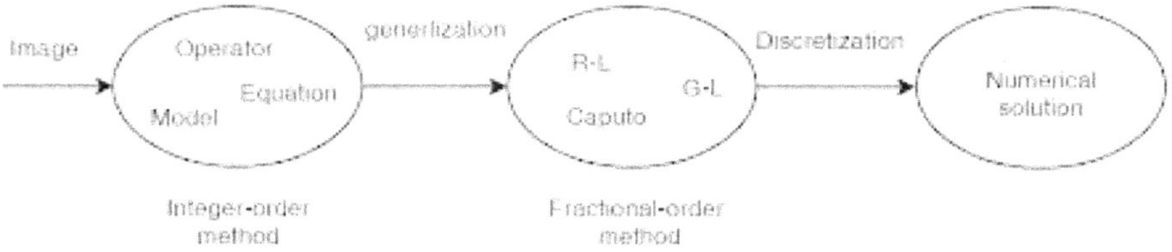

Fig. 8.1 Schematic diagram of application of fractional calculus.

processing, fractional order system is utilized for processing images and its systematic, diagram is shown below in Figure 8.1. In a few decades, many operators based on fractional calculus have been utilized in the image processing and invoking AI image processing techniques such as quality and texture enhancement, denoising, segmentation, restoration and impainting. No image will be vividly visible to human eye, if they don't have edges. Hence each image has at least some amount of edge information to discriminate an object from their background [45]. Regular edge detection method utilizes gradient convolution mask for the extraction of boundaries, which are easy to compute and incorporate in a real time application to image edge detection field. These masks are of odd order and modifies the image according to the pixel under consideration [43].

A basic idea behind image processing is the pairing of frequency of light absorbed to grey values. Edges in the image are identified by variation of intensity, between each neighboring pixels. Large frequency modulation is related to high difference in intensity values in an image. Similarly, small frequency modulation is related to a low difference in intensity values in an image. High modulation in intensity values denotes the presence of noise or edge of the image at that instance. Low modulation in intensity values denotes the background of the image at that instance [32].

Numerous approaches based on regular and fractional derivatives which have been coined by many researchers in [40–42]. In recent years, authors suggested the use of fractional derivative for processing images [18–20, 55, 70]. An improved condition for identifying edges are presented in detail in this section. Grunwald-Letnikov defined an non-integer order derivative of a continuous function as an extension for classical integer order derivative. Let for $\forall p \in R$, $I(s) \in [a, s]$ with $a < s, a, s \in R$ be a $q \in Z$ order continuously differentiable function. The non-integer order derivative of order q, where $p > \lceil q \rceil$ is defined as

$$ {}_aD_s^q I(s) = \lim_{h \to 0} \sum_{q=0}^{\lceil \frac{s-a}{h} \rceil} \frac{(-1)^p}{h^q} \frac{\Gamma(q+1)}{\Gamma(p+1)\Gamma(q-p+1)} I(s-ph) \tag{8.8} $$

The equation (8.8) can be represented as a difference equation for q^{th} order derivative as

$$ \begin{aligned} \frac{d^q}{ds^q} I(s) &= I(s) + (-q)I(s-1) + \frac{(-q)(-q+1)}{2!} I(s-2) \\ &+ \frac{(-q)(-q+1)(-q+2)}{3!} I(s-3) + \dots \\ &+ \frac{\Gamma(n-q)}{\Gamma(n+1)\Gamma(-q)} I(s-n) + \dots \end{aligned} $$

As in the case of one dimensional difference equation, q^{th} order fractional derivative for two dimensional function $I(u, v)$ can be expressed as:

$$\begin{aligned}\frac{d^q}{du^q}I(u,v) &= I(u,v) + \frac{(-q)I(u-1,v)}{1!} + \frac{(-q)(-q+1)}{2!}I(u-2,v)\\ &+\frac{(-q)(-q+1)(-q+2)}{3!}I(u-3,v) + \ldots\\ &+\frac{\Gamma(n-q)}{\Gamma(n+1)\Gamma(-q)}I(u-n,v) + \ldots \end{aligned} \tag{8.9}$$

$$\begin{aligned}\frac{d^q}{dv^q}I(u,v) &= I(u,v) + \frac{(-q)I(u,v-1)}{1!} + \frac{(-q)(-q+1)}{2!}I(u,v-2)\\ &+\frac{(-q)(-q+1)(-q+2)}{3!}I(u,v-3) + \ldots\\ &+\frac{\Gamma(n-q)}{\Gamma(n+1)\Gamma(-q)}I(u,v-n) + \ldots \end{aligned} \tag{8.10}$$

It is clear that in the n non-zero coefficients of the above equation (8.9) and equation (8.10), only first term is constant and the rest $(n-1)$ are coefficients are fractional order functions. Fractional mask with order v have been provided in Figure 8.2.

8.2.5 Generalized Fractal Dimensions for Experimental Images

In this section, we invoke the multifractal analysis in the recognition of noise free images by using Generalized Fractal Dimensions (GFD) [78]–[90].

Now we define a probability distribution of a experimental gray scale image by the following construction.

Let N be the number of boxes to cover the tested gray scale image with box size r. The probability p_i for i^{th} box of size r in the gray scale image is defined as,

$$p_i = \frac{M_i}{M}$$

where M_i is the mass of the tested gray scale image included in the corresponding i^{th} box of size r and M is the total mass of the tested gray scale image.

Then, the *Renyi Fractal Dimensions or Generalized Fractal Dimensions (GFD)* of order $q \in (-\infty, \infty)$ such that $q \neq 1$ of the tested gray scale image for the known probability distribution can be defined as

$$D_q = \lim_{r\to 0}\frac{1}{q-1}\frac{log_2\left(\sum_{i=1}^{N} p_i^q\right)}{log_2 r} \tag{8.11}$$

Here D_q is defined in terms of generalized Renyi Entropy.

8.2.5.1 Some Special Cases of Generalized Fractal Dimensions for Gray Scale Images

- If $q = 0$, then

$$D_0 = -\frac{log_2 N}{log_2 r}$$

which is nothing but the *Fractal Dimension* of the gray scale image.

⋯	0	$\frac{\Gamma(-v+1)}{(n-1)!\Gamma(-v+n)}$	0	⋯
⋯	⋮	⋮	⋮	⋯
⋯	0	$\frac{(-v)(-v+1)}{2}$	0	⋯
⋯	0	$-v$	0	⋯
⋯	0	1	0	⋯

(a)

⋮	⋮	⋮	⋮	⋮
0	⋯	0	0	0
$\frac{\Gamma(-v+1)}{(n-1)!\Gamma(-v+n)}$	⋯	$\frac{(-v)(-v+1)}{2}$	$-v$	1
0	⋯	0	0	0
⋮	⋮	⋮	⋮	⋮

(b)

⋯	0	1	0	⋯
⋯	0	$-v$	0	⋯
⋯	0	$\frac{(-v)(-v+1)}{2}$	0	⋯
⋯	⋮	⋮	⋮	⋯
⋯	0	$\frac{\Gamma(-v+1)}{(n-1)!\Gamma(-v+n)}$	0	⋯

(c)

⋮	⋮	⋮	⋮	⋮
0	0	0	⋯	0
1	$-v$	$\frac{(-v)(-v+1)}{2}$	⋯	$\frac{\Gamma(-v+1)}{(n-1)!\Gamma(-v+n)}$
0	0	0	⋯	0
⋮	⋮	⋮	⋮	⋮

(d)

⋯	0	0	0	1
⋯	0	0	$-v$	0
⋯	0	$\frac{(-v)(-v+1)}{2}$	0	0
0	⋰	0	0	0
$\frac{\Gamma(-v+1)}{(n-1)!\Gamma(-v+n)}$	0	⋮	⋮	⋮

(e)

⋮	⋮	⋮	0	$\frac{\Gamma(-v+1)}{(n-1)!\Gamma(-v+n)}$
0	0	0	⋰	0
0	0	$\frac{(-v)(-v+1)}{2}$	0	⋯
0	$-v$	0	0	⋯
1	0	0	0	⋯

(f)

$\frac{\Gamma(-v+1)}{(n-1)!\Gamma(-v+n)}$	0	⋮	⋮	⋮
0	⋱	0	0	0
⋯	0	$\frac{(-v)(-v+1)}{2}$	0	0
⋯	0	0	$-v$	0
⋯	0	0	0	1

(g)

1	0	0	0	⋯
0	$-v$	0	0	⋯
0	0	$\frac{(-v)(-v+1)}{2}$	0	⋯
0	0	0	⋱	0
⋮	⋮	⋮	0	$\frac{\Gamma(-v+1)}{(n-1)!\Gamma(-v+n)}$

(h)

Fig. 8.2 Eight directional fractional mask.

- As $q \longrightarrow 1$, D_q converges to D_1, which is given by

$$D_1 = \lim_{r \to 0} \frac{\sum_{i=1}^{N} p_i log_2 p_i}{log_2 r}.$$

This is called as *Information Dimension* of the gray scale image.
- If $q = 2$, then D_q is called the *Correlation Dimension* of the gray scale image.
- There are two limit cases of the gray scale image when $q = -\infty$ and $q = \infty$, which is given as

$$D_{-\infty} = \lim_{r \to 0} \frac{log_2 (p_{min})}{log_2 r}$$
$$D_{\infty} = \lim_{r \to 0} \frac{log_2 (p_{max})}{log_2 r}$$

where

$$p_{min} = \min\{p_1, p_2, ..., p_N\}$$
$$p_{max} = \max\{p_1, p_2, ..., p_N\}.$$

8.3 Proposed Method of Edge Detection

This section explains the working mechanism of the proposed method.

8.3.1 Image in Intuitionistic Fuzzy Domain

Image processing based on FS theory is introduced to address the problem of vagueness in image properties such as brightness, edges by patterning membership function. Though FS removes such uncertainty, there arises hesitation during the allotment of quantitative value of brightness to the considered pixel. The principal aim of the extended fuzzy image processing is to reduce gray level vagueness along with the elimination of ambiguity in the allotment membership values to those uncertain image pixels. These reasons make one to transfer an image in a FS domain to an IFS domain. Usually, experts may opt or define membership functions in an intuitive way. Therefore, vagueness in images ought to be reduced by considering a membership function without deciding the best among the choice of membership functions. The above discussion motivates the introduction of IFS in images to reduce the hesitation in assigning values to the brightness levels due to lack of knowledge/personal error.

The membership degree of the image I of size $P \times Q$ in IFS domain is calculated as

$$\mu_A(I(u, v)) = 1 - (1 - \mu_F(I(u, v)))^{\gamma}, \ \gamma \geq 0,$$

where the value of μ_F is obtained using equation

$$\mu_F(I(u, v)) = \frac{I(u, v) - g_{min}}{g_{max} - g_{min}},$$

where g_max and g_{min} is maximum and minimum intensity level in the image $I(u, v)$. The non-membership degree is computed as

$$\nu_A(I(u,v)) = \frac{1-\mu_A(I(u,v))}{1+\gamma.\mu_A(I(u,v))}.$$

The hesitation degree is defined as

$$\pi_A(I(u,v)) = 1-\mu_A(I(u,v)) - \nu_A(I(u,v)).$$

The image I in IFS domain is described as

$$\begin{cases} I_A = \left\{\langle I(u,v), \mu_A(I(u,v)), \nu_A(I(u,v)), \pi_A(I(u,v))\rangle\right\},\ 0 \le I(u,v) \le L-1, \\ 0 \le \mu_A(I(u,v)) \le 1, 0 \le \nu_A(I(u,v)) \le 1, 0 \le \pi_A(I(u,v)) \le 1,\ 1 \le u \le P,\ \ 1 \le v \le Q. \end{cases} \tag{8.12}$$

8.3.2 Steps of the Proposed Method of Edge Detection

Steps of the edge detection based on the fractional order derivative by enhancing image in IFS domain is as follows.

Step 1: Initially, image is converted into an image in IFS domain by utilizing the equation (8.12).
Step 2: The image is enhanced by using fuzzy enhancement operator *(Enh)* defined as follows

$$Enh(I(u,v)) = \begin{cases} 2I(u,v)^{\alpha}, if \ \ 0 \le I(u,v) < 0.5; \\ 1-2I(u,v)^{\alpha}, if \ \ 0.5 \le I(u,v) < 1. \end{cases} \tag{8.13}$$

Here α, $\alpha > 0$ in equation (8.13) is optimized by generating images for each α and the best enhancement image is opted by choosing image with average visibility. Visibility measure *(Vis)* is defined as

$$Vis = \sum_{i=1}^{M}\sum_{i=1}^{N}\left(I(u,v) - \frac{M_n}{M_n^{\beta+1}}\right),$$

where M_n denotes mean intensity of the image I, $\beta = 0.6$ to 0.7.
Step 3: Enhanced image from step 2 is defuzzified and is then convolved by fractional differential masks given in Figure 8.2 with $p > 0$.
Step 4: The order of fractional mask v is optimized by comparing generated edge images by fuzzy divergence

$$D(I) = \sum\sum 2-(2-\mu_A(I(u,v)))e^{\mu_A(I(u,v))-1} - \mu_A(I(u,v))e^{1-\mu_A(I(u,v))}$$

Among those images, an image having minimum divergence is opted as an best edge image.
Step 5: Comparison of GFD of the performance of the proposed method graphically with the existing methods.

8.4 Experimental Results and Discussion

Tests are performed on images of large databases which includes standard images, nutrient deficient leaf images, blood cell images and so on. The results of proposed methods are compared with Sobel, Prewitt,

Roberts and Canny edge detection methods. Edge detection results of Cameraman, Lena and Pears image have been depicted in Figure 8.3. It is clearly seen that the proposed method detects edges vividly than the other four classical methods in all the three standard images. Other methods have more under detected edges and over detected images than the proposed method.

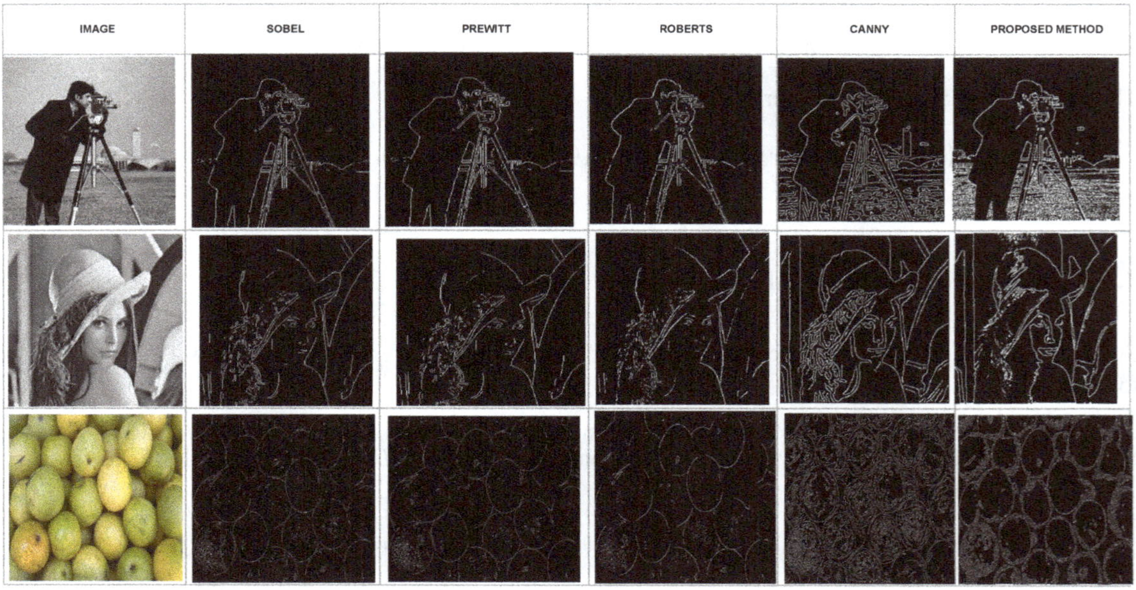

Fig. 8.3 Edge detection results of standard images.

Accuracy metrics have been evaluated for the edges of standard images and their values are presented as a bar diagram in Figure 8.4. In this bar diagram, for all the three images, the proposed method are relatively high when compared to other existing technique. Edge detection results of two nutrient deficiency leaf images and blood cell image are pictured in Figure 8.5. In these images, the proposed method seems to identify nutrient deficiency in leaf images and detects nuclei in blood cell images. The order, p of the fractional derivative for the 6 images are provided in the Table 8.1 below.

Table 8.1 Order of the fractional derivative for various images.

Image	ν
Camera Man	1.1
Lena	1.1
Pears	1.5
Cucumber Plant Leaf	1.8
Canola Plant Leaf	1.7
White Blood Cells	1.9

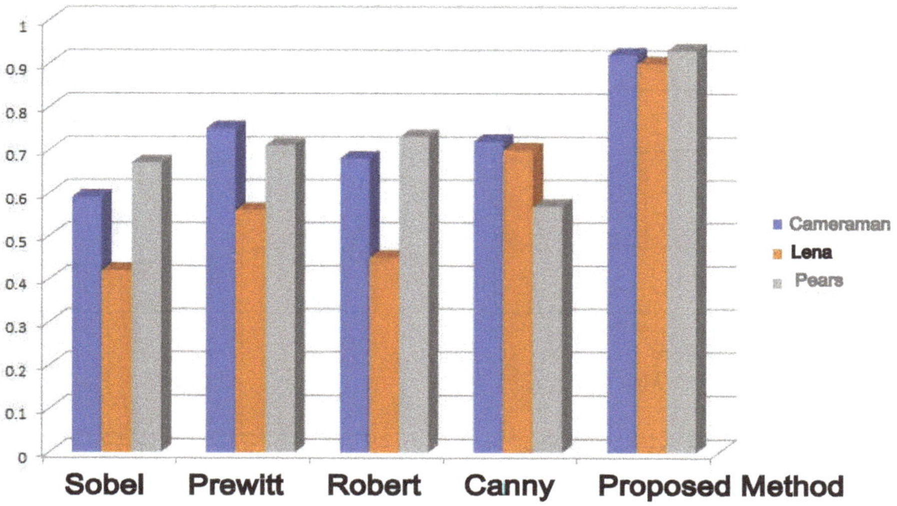

Fig. 8.4 Accuracy values.

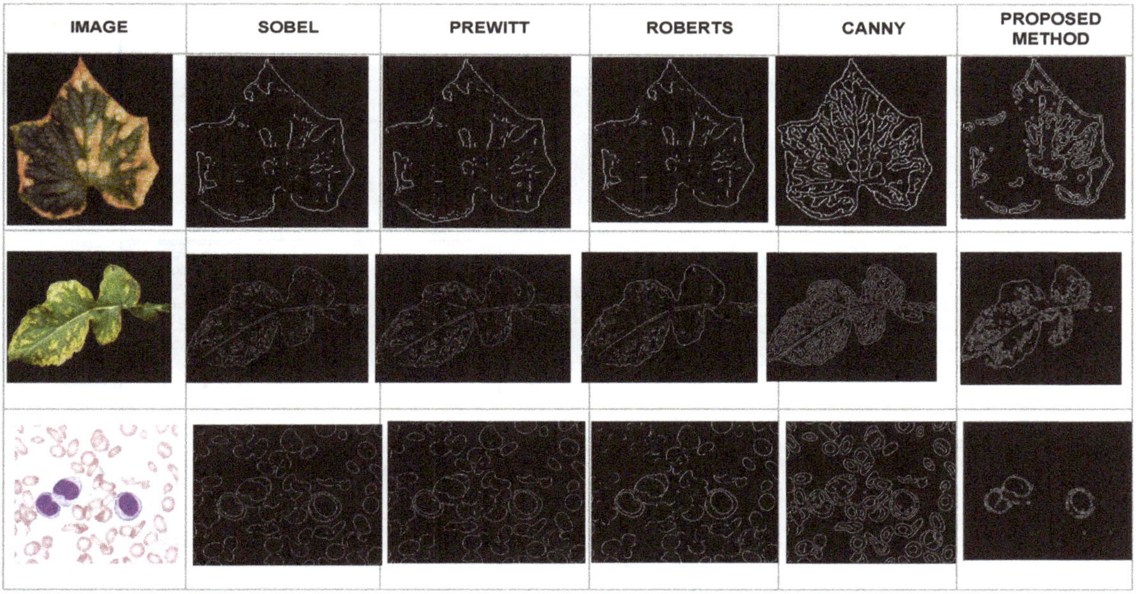

Fig. 8.5 Edge detection results of images with deficiency.

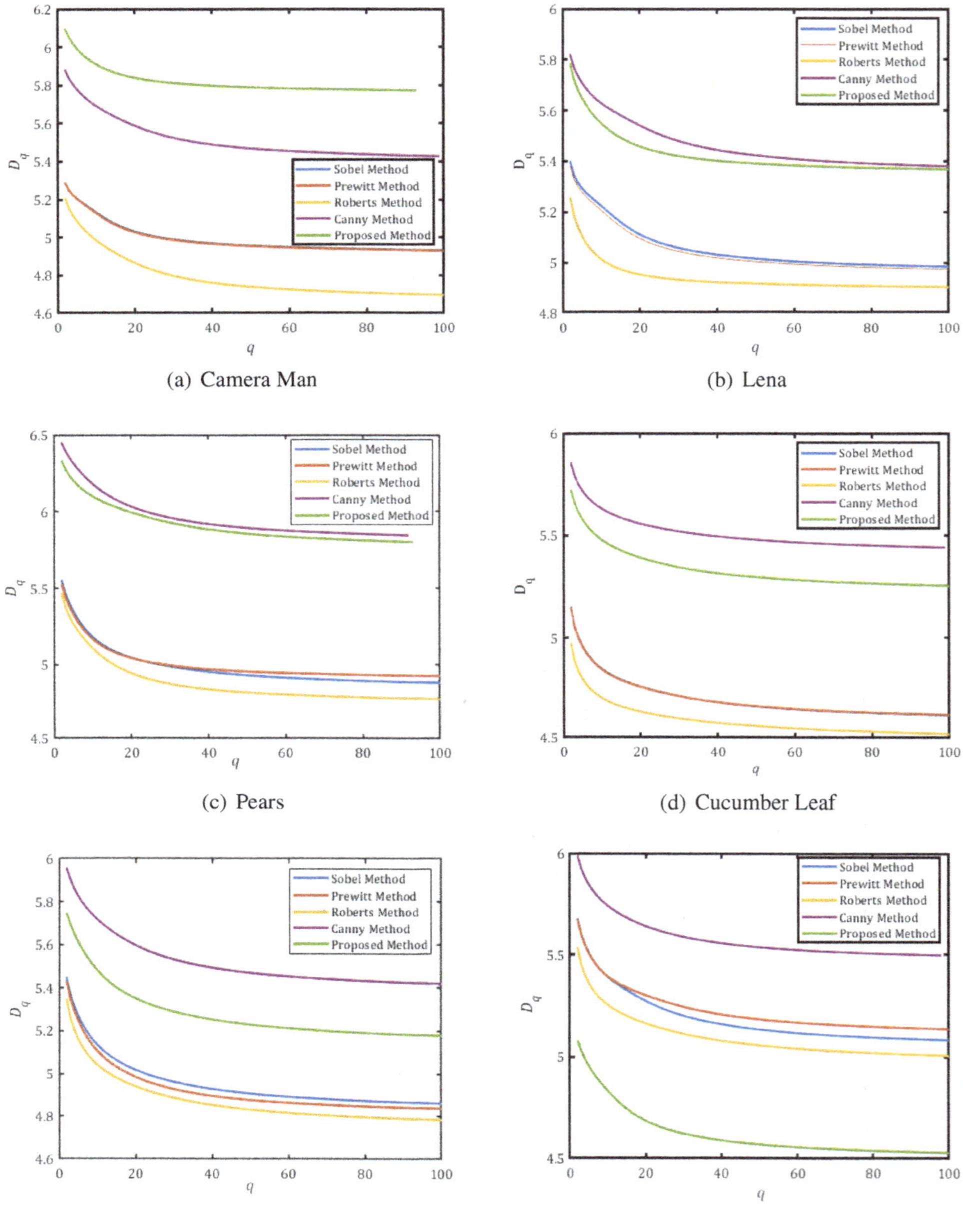

(a) Camera Man (b) Lena

(c) Pears (d) Cucumber Leaf

(e) Canola Leaf (f) Blood Cells

Fig. 8.6 Comparison of generalized fractal dimensions for the proposed method with the existing edge detection methods for various images.

It is clearly seen from the table that the values for standard images are from 1.1 to 1.5 and the blood cell and nutrient deficient leaf images are between 1.5 to 1.9. The results of the proposed methods for the standard images were observed to be improved from the result of the comparable methods. But, in case of nutrient deficient leaf images, the edges are detected by the proposed method is a simple extraction of deficient portion. In case of the white blood cell image, the proposed method extracts nucleus region. This may be due the order range between 1 and 2.

In Figure 8.6, it is clearly observed that the GFD for the proposed method is higher in the Camera Man image, compared to the other methods. It projects that the proposed method is employed with more complexity in the edge detection process in Camera Man image. Consequently the proposed method is significantly higher than other methods except the Canny method in Lena, Pears, Cucumber Plant Leaf and Canola Plant Leaf; whereas the proposed method performs the edge detection process with less complexity in Blood Cells images, as GFD of the same is lesser than all other methods.

8.5 Concluding Remarks

This chapter provides a new approach to the detection of edges in various standard images. The proposed technique works as an nutrient deficiency detector in leaf images and extracts nucleus in blood cells. The edges are detected in complex images tested in this research work by the proposed scheme, better than the comparable methods. The multifractal dimensional measures also supported the efficiency of performance of the proposed method by comparing with the existing methods. It is observed that the improved detection is due to the absorption of uncertainty in choosing the order of the fractional derivative for the tested images. As an extension of this research work, the extraction of nutrient deficiency in leaf images, and also the detection of tumors in brain and breast images can be done.

References

[1] Szmidt, E. and J. Kacprzyk. 2000. Distances between intuitionistic fuzzy sets. Fuzzy Sets and Systems, 114(3): 505–518.

[2] Atanassov, K.T. 1999. Intuitionistic Fuzzy Sets. Springer, 1–137.

[3] Zadeh, A.L. 1965. Information and control. Fuzzy Sets, 8(3): 338–353.

[4] Kumar, S., R. Saxena and K. Singh. 2017. Fractional Fourier transform and fractional-order calculus-based image edge detection. Circuits, Systems, and Signal Processing, 36(4): 1493–1513.

[5] Guan, J., J. Ou, Z. Lai and Y. Lai. 2018. Medical image enhancement method based on the fractional order derivative and the directional derivative. International Journal of Pattern Recognition and Artificial Intelligence, 32(03): 1857001.

[6] Ansari, M.D., A.R. Mishra and F.T. Ansari. 2018. New divergence and entropy measures for intuitionistic fuzzy sets on edge detection. International Journal of Fuzzy Systems, 20(2): 474–487.

[7] Atanassov, K.T. 1989. More on intuitionistic fuzzy sets. Fuzzy Sets and Systems, 33(1): 37–45.

[8] Chaira, T. and A.K. Ray. 2008. A new measure using intuitionistic fuzzy set theory and its application to edge detection. Applied Soft Computing, 8(2): 919–927.

[9] Vlachos, I.K. and G.D. Sergiadis. 2007. Intuitionistic fuzzy information–applications to pattern recognition. Pattern Recognition Letters, 28(2): 197–206.

[10] Mao, J., D. Yao and C. Wang. 2013. A novel cross-entropy and entropy measures of IFSs and their applications. Knowledge-Based Systems, 48: 37–45.
[11] Wu, D. and J.M. Mendel. 2010. On the continuity of type-1 and interval type-2 fuzzy logic systems. IEEE Transactions on Fuzzy Systems, 19(1): 179–192.
[12] Melin, P., C.I. Gonzalez, J.R. Castro, O. Mendoza and O. Castillo. 2014. Edge-detection method for image processing based on generalized type-2 fuzzy logic. IEEE Transactions on Fuzzy Systems, 22(6): 1515–1525.
[13] Bhardwaj, S. and A. Mittal. 2012. A survey on various edge detector techniques. Procedia Technology, 4: 220–226.
[14] Bustince, H., E. Barrenechea, M. Pagola, Miguel and J. Fernández. 2009. Interval-valued fuzzy sets constructed from matrices: Application to edge detection. Fuzzy Sets and Systems, 160(13): 1819–1840.
[15] Biswas, R. and J. Sil. 2012. An improved canny edge detection algorithm based on type-2 fuzzy sets. Procedia Technology, 4: 820–824.
[16] Pu, Y.F., J.L. Zhou and X. Yuan. 2010. Fractional differential mask: A fractional differential-based approach for multiscale texture enhancement. IEEE Transactions on Image Processing, 19(2): 491.
[17] Mathieu, B., P. Melchior, A. Oustaloup and C.H. Ceyral. 2003. Fractional differentiation for edge detection. Signal Processing, 83(11): 2421–2432.
[18] Gao, C., J. Zhou, X. Zheng and F. Lang. 2011. Image enhancement based on improved fractional differentiation. Journal of Computational Information Systems, 7(1): 257–264.
[19] Gao, C.B., J.L. Zhou, J.R. Hu and F.N. Lang. 2011. Edge detection of colour image based on quaternion fractional differential. IET Image Processing, 5(3): 261–272.
[20] Gao, C.B. and J.L. Zhou. 2010. Image enhancement based on quaternion fractional directional differentiation. Acta Automatica Sinica, 37(2): 50–159.
[21] Caputo, M. 1967. Linear models of dissipation whose Q is almost frequency independent II. Geophysical Journal International, 13(5): 529–539.
[22] Watanabe, Y. 1931. Notes on the generalized derivatives of Riemann-Liouville and its applications to Leibniz formula I and II. Tohoku Math. J., 34: 8–14.
[23] Marchaud, A. 1927. On the derivatives and on the differences of the functions of real variables, impr. Gauthier-Villars.
[24] Weyl, H. 1917. Remarks on the concept of the differential quotient of the fractional order, quarter jschr. Natural research. Social Zurich, 62(1-2): 296–302.
[25] Letnikov, A.V. 1868. Theory of differentiation with an Arbtraly indicator. Matem Sbornik, 3: 1–68.
[26] Grunwald, A.K. 1867. About limited derivatives and their application. Zangew Math Und Phys., 12: 441–480.
[27] Laurent, H. 1884. On the calculation of the revised indexes. Nouvelles Annales de Mathe Matiques: Journal of Candidates for Polytechnic and Normal Schools, 3: 240–252.
[28] Sonin, N.Y. 1869. On differentiation with arbitrary index. Moscow Matem. Sbornik, 6(1): 1–38.
[29] Letnikov, A.V. 1872. An explanation of the concepts of the theory of differentiation of arbitrary index (Russian). Moscow Matem, Sbornik, 6: 413–445.
[30] Riemann, B. Attempt at a general understanding of integration and differentiation. Collected Works, 62(1876).
[31] Cayley, A. 1880. Note on Riemann's paper Versuch einer allgemeinen Auffassung der Integration und Differentiation. Mathematische Annalen, 16(1): 81–82.
[32] McAndrew, A. 2004. An introduction to digital image processing with matlab notes for scm2511 image processing. School of Computer Science and Mathematics, Victoria University of Technology, 264(1): 1–264.

[33] Liouville, J. 1832. Memory on some questions of geometry and mechanics, and on a new kind of calculus for reweld these questions.

[34] Liouville, J. 1832. Memo on the integration of the equation $(mx2+nx+p)d2y/dx2+(qx+pr)dy/dx+sy = 0$ using the differentials with arbitrary indices. Journal d l'Ecole Polytechnique. 13(1832): 163–186.

[35] Liouville, J. 1832. Memoir on some questions of geometry and mechanics, and on a new kind of calculation to solve these questions. J. de l'Ecole Pol. Tech, 13(1832): 1–69.

[36] Baron Fourier, J.B.J. 1822. The analytical orie of heat, (Chez Firmin Didot, father and son).

[37] Lacroix, S.F. 1797. Trait'e of differential calculus and integral calculus, 1 (Trait'e of differential calculus and integral calculus).

[38] Lagrange, J.L. 1772. Sur une nouvelle espece de calcul relatif ala differentiation et al integration des quantites variables. Ouvres de Lagrange.

[39] Euler, L. 1738. On transcendental progressions or whose general terms cannot be given algebraically. Commentaries of the University of Petersburg, 36–57.

[40] Gao, C., J. Zhou and W.H. Zhang. 2014. Edge detection based on the newton interpolation's fractional differentiation. Int. Arab J. Inf. Technol., 11(3): 223–228.

[41] Yang, Z., F. Lang, X. Yu and Y.U. Zhang. 2011. The construction of fractional differential gradient operator. Journal of Computational Information Systems, 7(12): 4328–4342.

[42] Oustaloup, A., B. Mathieu and P. Melchior. 1991. Edge detection using non integer derivation. IEEE European Conference on Circuit Theory and Design (ECCTD91), Copenhagen, Denmark, 3–6.

[43] Jalab, H.A. and R.W. Ibrahim. 2013. Texture enhancement based on the Savitzky-Golay fractional differential operator. Mathematical Problems in Engineering, 2013.

[44] Jalab, H.A. and R.W. Ibrahim. 2015. Fractional Alexander polynomials for image denoising. Signal Processing, 107: 340–354.

[45] Appati, J.K. 2017. Construction of a novel convolution based fractional derivative mask for image edge analysis.

[46] Petras, I. 2011. Fractional derivatives, fractional integrals, and fractional differential equations in Matlab, (IntechOpen).

[47] Dingyu, X. 2006. Control System Computer Aided Design-MATLAB Language and Application, Beijing: Publishing House of Tsinghua University.

[48] Magin, R.L. 2006. Fractional calculus in bioengineering, 2(6).

[49] Magin, R.L., X.U. Feng and D. Baleanu. 2008. IFAC Proceedings Volumes, 41(2): 9613–9618.

[50] Chen, Y.Q. and K.L. Moore. 2002. Discretization schemes for fractional-order differentiators and integrators. IEEE Transactions on Circuits and Systems I: Fundamental Theory and Applications, 49(3): 363–367.

[51] Magin, R.L. and M. Ovadia. 2008. Modeling the cardiac tissue electrode interface using fractional calculus. Journal of Vibration and Control, 14(9-10): 1431–1442.

[52] Magin, R.L., O. Abdullah, D. Baleanu and X.J. Zhou. 2008. Anomalous diffusion expressed through fractional order differential operators in the Bloch–Torrey equation. Journal of Magnetic Resonance, 190(2): 255–270.

[53] Oustaloup, A., F. Levron, B. Mathieu and F.M. Nanot. 2000. Frequency-band complex noninteger differentiator: Characterization and synthesis. IEEE Transactions on Circuits and Systems I: Fundamental Theory and Applications, 47(1): 25–39.

[54] Frangi, A.F., W.J. Niessen, K.L. Vincken and M.A. Viergever. 1998. Multiscale vessel enhancement filtering. International Conference on Medical Image Computing and Computer-Assisted Intervention, 130–137.

[55] Mathieu, B., P. Melchior, A. Oustaloup and C. Ceyral. 2003. Fractional differentiation for edge detection. Signal Processing, 83(11): 2421–2432.

[56] Podlubny, I. 1999. Mathematics in Science and Engineering 198. Fractional Differential Equations.

[57] Shi, J. and J. Malik. 2000. Normalized cuts and image segmentation. IEEE Transactions on Pattern Analysis and Machine Intelligence, 22(8): 888–905.

[58] Li, C., C.Y. Kao, J.C. Gore and Z. Ding. 2008. Minimization of region-scalable fitting energy for image segmentation. IEEE Transactions on Image Processing: A Publication of the IEEE Signal Processing Society, 17(10): 1940.

[59] Martínez-Jiménez, L., J.M. Cruz-Duarte, J.J. Rosales and I. Cruz-Aceves. 2018. Enhancement of vessels in coronary angiograms using a Hessian matrix based on Grunwald-Letnikov fractional derivative. Proceedings of the 2018 8th International Conference on Biomedical Engineering and Technology, 51–54.

[60] He, N., J.B. Wang, L.L. Zhang and K. Lu. 2015. An improved fractional-order differentiation model for image denoising. Signal Processing, 112: 180–188.

[61] Otsu, N. 1979. A threshold selection method from gray-level histograms. IEEE Transactions on Systems, Man, and Cybernetics, 9(1): 62–66.

[62] Pu, Y., W. Wang, J. Zhou, Y. Wang and H. Jia. 2008. Fractional differential approach to detecting textural features of digital image and its fractional differential filter implementation. Science in China Series F: Information Sciences, 51(9): 1319–1339.

[63] Yang, Q., D. Chen, T. Zhao and Y. Chen. 2016. Fractional calculus in image processing: A review. Fractional Calculus and Applied Analysis, 19(5): 1222–1249.

[64] Rashmi, M. Kumar and R. Saxena. 2013. Algorithm and technique on various edge detection: A survey. Signal & Image Processing, 4(3): 65.

[65] Sarkar, S., S. Paul, R. Burman, S. Das and S.S. Chaudhuri. 2014. A fuzzy entropy based multi-level image thresholding using differential evolution. International Conference on Swarm, Evolutionary, and Memetic Computing, 386–395.

[66] Suresh, S. and S. Lal. 2016. An efficient cuckoo search algorithm based multilevel thresholding for segmentation of satellite images using different objective functions. Expert Systems with Applications, 58: 184–209.

[67] Vicente, S., C. Rother and V. Kolmogorov. 2011. Object cosegmentation. Computer Vision and Pattern Recognition (CVPR), 2011 IEEE Conference, 2217–2224.

[68] Chan, T.F. and L.A. Vese. 2001. Active contours without edges. IEEE Transactions on Image Processing, 10(2): 184–209.

[69] Ananthi, V.P., P. Balasubramaniam and C.P. Lim. 2014. Segmentation of gray scale image based on intuitionistic fuzzy sets constructed from several membership functions. Pattern Recognition, 47(12): 3870–3880.

[70] Pu, Y.F., J.L. Zhou and X. Yuan. 2010. Fractional differential mask: A fractional differential-based approach for multiscale texture enhancement. IEEE Transactions on Image Processing, 19(2): 491–511.

[71] Mandelbrot, B.B. 1983. The Fractal Geometry of Nature, W.H. Freeman and Company, New York.

[72] Barnsley, M.F. 2014. Fractals Everywhere, Academic Press, USA.

[73] Barnsley, M.F. 2006. SuperFractals, Cambridge University Press, New York.

[74] Falconer, K. 2003. Fractal Geometry: Mathematical Foundations and Applications, John Wiley & Sons Ltd., England.

[75] Edgar, G. 2008. Measure, Topology, and Fractal Geometry, Springer, New York.

[76] Banerjee, S., M.K. Hassan, Sayan Mukherjee and A. Gowrisankar. 2019. Fractal Patterns in Nonlinear Dynamics and Applications, CRC Press.

[77] Banerjee, S., D. Easwaramoorthy and A. Gowrisankar. 2021. Fractal Functions, Dimensions and Signal Analysis, Understanding Complex Systems, Springer: Complexity, Springer, Cham.

[78] Easwaramoorthy, D. and R. Uthayakumar. 2010. Estimating the complexity of biomedical signals by multifractal analysis. Proceedings of the IEEE Students' Technology Symposium. IEEE Xplore Digital Library, IEEE, USA, 6–11.

[79] Easwaramoorthy, D. and R. Uthayakumar. 2010. Analysis of EEG signals using advanced generalized fractal dimensions. Proceedings of the Second International Conference on Computing, Communication and Networking Technologies, IEEE Xplore Digital Library, IEEE, USA, 1–6.

[80] Easwaramoorthy, D. and R. Uthayakumar. 2010. Analysis of biomedical EEG signals using wavelet transforms and multifractal analysis. Proceedings of the IEEE International Conference on Communication Control and Computing Technologies, IEEE Xplore Digital Library, IEEE, USA, 544–549.

[81] Easwaramoorthy, D. and R. Uthayakumar. 2011. Improved generalized fractal dimensions in the discrimination between healthy and epileptic EEG signals. Journal of Computational Science, 2(1): 31–38.

[82] Uthayakumar, R. and D. Easwaramoorthy. 2012. Multifractal-wavelet based denoising in the classification of healthy and epileptic EEG signals. Fluctuation and Noise Letters, 11(4): 1250034.

[83] Uthayakumar, R. and D. Easwaramoorthy. 2012. Generalized fractal dimensions in the recognition of noise free images. Proceedings of the International Conference on Computing, Communication and Networking Technologies, IEEE Xplore Digital Library, IEEE, USA, 1–5.

[84] Uthayakumar, R. and D. Easwaramoorthy. 2012. Multifractal analysis in denoising of color images. Proceedings of the International Conference on Emerging Trends in Science, Engineering and Technology, IEEE Xplore Digital Library, IEEE, USA, 228–234.

[85] Uthayakumar, R. and D. Easwaramoorthy. 2013. Epileptic seizure detection in EEG signals using multifractal analysis and wavelet transform. Fractals, 21(2): 1350011.

[86] Uthayakumar, R. and D. Easwaramoorthy. 2014. Fuzzy generalized fractal dimensions for chaotic waveforms. Chaos, Complexity and Leadership 2012, Springer Proceedings in Complexity, 411–422.

[87] Easwaramoorthy, D., P.S. Eliahim Jeevaraj, A. Gowrisankar, A. Manimaran and S. Nandhini. 2018. Fuzzy generalized fractal dimensions using inter-heartbeat interval dynamics in ECG signals for age related discrimination. International Journal of Engineering and Technology (UAE), 7(4.10): 900–903.

[88] Nandhini, S., D. Easwaramoorthy and R. Abinands. 2020. An extensive review on recent evolutions in object detection algorithms. International Journal of Emerging Trends in Engineering Research, 8(7): 3766–3776.

[89] Priyanka, T.M.C. and A. Gowrisankar. 2021. Analysis on Weyl-Marchaud Fractional Derivative for Types of Fractal Interpolation Function with Fractal Dimension, Fractals.

[90] Easwaramoorthy, D., A. Gowrisankar, A. Manimaran, S. Nandhini, L. Rondoni and S. Banerjee. 2021. An exploration of fractal-based prognostic model and comparative analysis for second wave of COVID-19 diffusion. Nonlinear Dynamics.

For Product Safety Concerns and Information please contact our EU
representative GPSR@taylorandfrancis.com
Taylor & Francis Verlag GmbH, Kaufingerstraße 24, 80331 München, Germany

www.ingramcontent.com/pod-product-compliance
Lightning Source LLC
Chambersburg PA
CBHW081530220326
41598CB00036B/6390
9781032138732